Letts

EDUCATIONAL

ADVANCED LEVEL

Revise A2

Physics

Author

Graham Booth

Contents

Specification lists

AQA A Physics

MODULE	SPECIFICATION TOPIC	CHAPTER REFERENCE	STUDIED IN CLASS	REVISED	PRACTICE QUESTIONS
Module 4 (M4) *Waves, fields and nuclear energy*	Oscillations and waves	2.1, 2.2, 2.3, 2.4			
	Capacitance	3.3			
	Gravitational and electric fields	3.1, 3.2, 3.5			
	Magnetic effects of currents	3.4, 3.5, 3.6			
	Nuclear applications	4.3			
Module 5 (M5) *Nuclear Instability*	Nuclear instability	4.2, 4.4			
Module 6A (M6A) *Astrophysics*	Lenses and optical telescopes	6.1			
	Radio astronomy	6.1			
	Classification of stars	6.2, 6.4			
	Cosmology	6.3, 6.4			
Module 6B (M6B) *Medical physics*	Physics of the eye and ear	5.2, 5.3			
	Biological measurement and imaging	5.4			

Examination analysis

The specification comprises three compulsory modules. In module tests all the questions are compulsory.

Module 4	15 compulsory multiple choice questions and several short, structured questions	1 hr 30 min test	30%
Module 5	Structured questions on module 5 and one option topic **and** coursework **or** practical examination	1 hr 15 min test 1 hr 30 min practical	30%
Module 6	Synoptic test Structured questions on modules 1–5	2 hr test	40%

AQA B Physics

MODULE	SPECIFICATION TOPIC	CHAPTER REFERENCE	STUDIED IN CLASS	REVISED	PRACTICE QUESTIONS
Module 4 (M4) *Further Physics*	Circular motion	1.4			
	Oscillations	2.1, 2.2			
	Work and energy	1.3			
	Molecular kinetic theory	4.1			
	Heating and working	4.1			
	Capacitance and exponential decay	3.3			
	Momentum concepts	1.2			
	Quantum phenomena	2.5			
Module 5 (M5) *Fields and their applications*	Electric and gravitational fields	3.1, 3.2, 7.1			
	Magnetic fields	4.3			
	Nuclear energy particle, accelerators and detectors	4.2, 7.2			

Examination analysis

The specification comprises three compulsory modules. In module tests all the questions are compulsory.

Module 4	Structured questions	1 hr 30 min test	30%
Module 5	Structured questions on module 5 and synoptic questions on modules 1–5	2 hr test	40%
Module 6	Practical examination	3 hr test	30%

Edexcel A Physics

MODULE	SPECIFICATION TOPIC	CHAPTER REFERENCE	STUDIED IN CLASS	REVISED	PRACTICE QUESTIONS
Module 4 (M4) *Waves and our Universe*	Circular motion and oscillations	1.4, 2.2			
	Waves	2.3, 2.5			
	Superposition of waves	2.4			
	Quantum phenomena	2.5			
	The expanding Universe	6.3			
Module 5 (M5) *Fields and forces*	Gravitational fields	3.1			
	Electric fields	3.2			
	Capacitance	3.3			
	Magnetic fields	3.4			
	Electromagnetic induction	3.6			
Module 6 (M6) *Synthesis*	Analogies in physics	7.1			
	Accelerators	3.5, 7.2			

Examination analysis

The specification comprises three compulsory modules. In module tests all the questions are compulsory.

Module 4	Structured questions	1 hr 20 min test	30%
Module 5	Practical examination	1 hr 30 min	15%
	Structured questions	1 hr test	15%
Module 6	Synoptic test 1 comprehension question 1 question based on module 6 2 questions based on any area except the options topics	2 hr test	40%

Edexcel B Physics

MODULE	SPECIFICATION TOPIC	CHAPTER REFERENCE	STUDIED IN CLASS	REVISED	PRACTICE QUESTIONS
Module 4 (M4) *Moving with physics*	Transport on track	1.2, 3.3, 3.4, 3.6			
	The medium is the message	3.2, 3.3, 3.4			
	Probing the heart of matter	1.2, 1.4, 2.5, 3.2, 4.3, 4.4, 7.2			
Module 5 (M5) *Physics from creation to collapse*	Reach for the stars	3.1, 4.1, 4.2, 4.3, 6.3, 6.4, 7.1			
	Build or bust?	2.1, 2.2			

Examination analysis

The specification comprises three compulsory modules. In module tests all the questions are compulsory.

Module 4	Structured and free-response questions	1 hr 30 min test	30%
Module 5	Structured and free-response questions	1 hr test	20%
	Coursework		20%
Module 6	Synoptic test 1 comprehension question Structured and free-response questions	1 hr 30 min test	30%

OCR A Physics

MODULE	SPECIFICATION TOPIC	CHAPTER REFERENCE	STUDIED IN CLASS	REVISED	PRACTICE QUESTIONS
Module 4 (M4) Forces, fields and energy	Dynamics	1.1, 1.2			
	Work and energy	1.3			
	Motion in a circle	1.4			
	Oscillations	2.1, 2.2			
	Gravitational fields	3.1			
	Electric fields	3.2			
	Capacitors	3.3			
	Electromagnetism	3.4, 3.5			
	Electromagnetic induction	3.6			
	Thermal physics	4.1			
	The nuclear atom	4.3, 4.4			
	Radioactivity	4.2			
Module 5.1 (M5.1) Cosmology	Models of the known Universe	6.2			
	Stars and galaxies	6.2			
	Structure of the Universe	6.2, 6.3			
	Information from stellar observation	6.3, 6.4			
	How the Universe may evolve	6.3			
	Relativity	6.5			
Module 5.2 (M5.2) Health physics	Body mechanics	5.1			
	The eye and sight	5.2			
	The ear and hearing	5.3			
	Medical imaging	5.4			
	Medical treatment	5.4			

Examination analysis

The specification comprises three compulsory modules. In module tests all the questions are compulsory.

Module 4	Structured and free-response questions	1 hr 30 min test	30%
Module 5	Test on one optional topic Structured and free-response questions	1 hr 30 min test	30%
Module 6	Synoptic test: questions on module 1–4	1 hr 15 min test	20%
	Coursework or practical examination	1 hr 30 min test	20%

OCR B Physics

MODULE	SPECIFICATION TOPIC	CHAPTER REFERENCE	STUDIED IN CLASS	REVISED	PRACTICE QUESTIONS
Module 4 (M4) *Rise and fall of the Clockwork Universe*	Creating models	1.3, 2.1, 2.2, 3.3, 4.2			
	Out into space	1.2, 1.3, 3.1			
	Our place in the Universe	6.3			
	Matter: very simple	4.1			
	Matter: hot or cold	4.1			
Module 5 (M5) *Field and particle pictures*	Electromagnetic machines	3.4, 3.6			
	Charge and field	3.2			
	Probing deep into matter	2.5, 3.4, 4.4, 7.2			
	Ionising radiation and risk	4.2, 4.3			

Examination analysis

The specification comprises three compulsory modules. In module tests all the questions are compulsory. They consist of short focused questions and longer, structured questions with some questions being open-ended.

Module 4	Theory test	1 hr 10 min	18.4%
Module 5	Theory test	1 hr 10 min test	21.6%
Module 6	Synoptic test: comprehension/data analysis and structured questions from any area of the specification	1 hr 30 min	30%
	Coursework: practical investigation and research report		30%

WJEC Physics

MODULE	SPECIFICATION TOPIC	CHAPTER REFERENCE	STUDIED IN CLASS	REVISED	PRACTICE QUESTIONS
Module 4 (M4) *Oscillations and energy*	Vibrations	1.4, 2.1, 2.2			
	Momentum concepts	1.1, 1.2			
	Energy concepts	1.3			
	Molecular kinetic theory	4.1			
	Capacitance	3.3			
	Alternating currents	00			
Module 5 (M5) *Fields, forces and nuclei*	Uniform and radial fields of force	3.1, 3.2			
	B-fields	3.4			
	Electromagnetic induction	3.6			
	Radioactivity and radioisotopes	4.2, 4.3			
	Nuclear energy	4.3			
	Probing matter	4.4			

Examination analysis

The specification comprises three compulsory modules.

Module 4	Theory test	1 hr 30 min test	30%
Module 5	Theory test, including some synoptic assessment	1 hr 30 min test	30%
Module 6	Synoptic test: questions covering all areas of the specification coursework	2 hr test	25%
	Coursework: practical investigatory task		15%

NICCEA Physics

MODULE	SPECIFICATION TOPIC	CHAPTER REFERENCE	STUDIED IN CLASS	REVISED	PRACTICE QUESTIONS
Module 4 (M4) *Energy, oscillations and fields*	Momentum and energy	1.2, 1.3			
	Thermal physics	4.1			
	Uniform circular motion	1.4			
	Oscillations	2.1, 2.2			
	Fields	3.1, 3.2, 3.5			
Module 5 (M5) *Electromagnetism and nuclear physics*	Capacitors	3.3			
	Electromagnetism	3.4, 3.6			
	Electron physics	3.2, 3.4			
	Atomic and nuclear physics	4.2, 4.4			
Module 6 (M6) *Particle physics*	Particle physics	3.5, 4.3, 4.4, 7.2			

Examination analysis

The specification comprises three compulsory modules. In module tests, all questions are compulsory.

Module 4	Theory test: short-answer questions and a synoptic, data analysis question	*1 hr 30 min test*	30%
Module 5	Theory test: short-answer questions and a synoptic, comprehension	*1 hr 30 min test*	30%
Module 6	Theory test: short-answer questions and a synoptic, free-response question	*1 hr test*	16.6%
	Practical examination	*1 hr 30 min*	23.4%

AS/A2 Level Physics courses

AS and A2

All Physics A Level courses being studied from September 2000 are in two parts, with three separate units or modules in each part. Most students start by studying the AS (Advanced Subsidiary) course. Some then go on to study the second part of the A Level course, called the A2. It is also possible to study the full A Level course, both AS and A2, in any order.

How will you be tested?

Assessment units

For A2 Physics, you will be tested by three assessment units. These form 50% of the weighting for the full A Level, the remaining 50% being the assessment of AS.

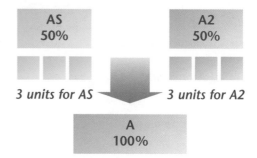

Each unit can normally be taken in either January or June. Alternatively, you can study the whole course before taking any of the unit tests. There is a lot of flexibility about when exams can be taken and the diagram below shows just some of the ways that the assessment units may be taken for AS and A Level Physics.

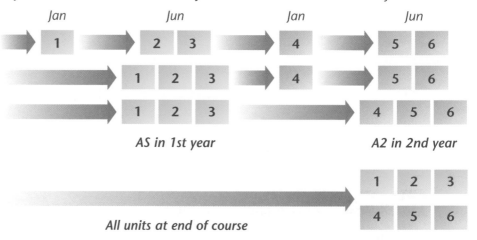

If you are disappointed with a module result, you can resit each module once. You will need to be very careful about when you take up a resit opportunity because you will have only one chance to improve your mark. The higher mark counts.

A2 and Synoptic assessment

All full A Level specifications have to include an element of synoptic assessment, one that draws together ideas from all areas of physics studied at AS and A2 Level. Details of these synoptic assessments are given in the specification lists. The synoptic assessment can examine any area of the specification except any optional topics. Questions will assess the extent to which you can apply your knowledge and understanding of physics from more than one course unit, for example a question set in the context of renewable energy could examine understanding of the motion of satellites (the Moon), simple harmonic motion (the movement of tides) and electromagnetic induction (electricity generation).

Synoptic assessment will often include comprehension and data handling exercises and require you to apply fundamental physical principles in unfamiliar situations. In some specifications the synoptic assessment is by a separate test, and in others it is built into all the A2 module tests. However it is done, it carries 30–40% of the marks for A2 and 15–20% of the total marks for the full A Level award.

Coursework

Coursework may form part of your A2 Level Physics course, depending on which specification you study. Where students have to undertake coursework, it may be a written project or for the assessment of practical skills or both. If in doubt, check with your teacher or the specification that you are studying.

Key Skills

To gain the key skills qualification, which is equivalent to an AS Level, you will need to demonstrate that you have attained level 3 in the areas of Communication, Application of number and Information technology. Part of the assessment can be done as normal class activity and part is by formal test.

What skills will I need?

For A2 Physics, you will be tested by assessment objectives: these are the skills and abilities that you should have acquired by studying the course. The assessment objectives for A2 Physics are shown below.

Knowledge with understanding

- recall of facts, terminology and relationships
- understanding of principles and concepts
- drawing on existing knowledge to show understanding of the responsible use of physics in society
- selecting, organising and presenting information clearly and logically.

Application of knowledge and understanding, analysis and evaluation

- explaining and interpreting principles and concepts
- interpreting and translating, from one form into another, data presented as continuous prose or in tables, diagrams and graphs
- carrying out relevant calculations
- applying knowledge and understanding to familiar and unfamiliar situations
- assessing the validity of chemical information, experiments, inferences and statements

You must also present arguments and ideas clearly and logically, using specialist vocabulary where appropriate.

Experimental and investigative skills

Physics is a practical subject and part of the assessment of AS Physics will test your practical skills. This may be done during your lessons or you may be tested in a more formal practical examination. You will be assessed on four main skills:

- planning
- implementing
- analysing evidence and drawing conclusions
- evaluating evidence and procedures

Different types of questions in A2 examinations

In order to assess these abilities and skills, a number of different types of question are used.

In A2 Level Physics, unit tests include short-answer questions, structured questions requiring both short answers and more extended answers, together with free-response and open-ended questions. Multiple choice question papers are not used, although it is possible that some short-answer questions will use a multiple choice format, where you have to choose the correct response from a number of given alternatives.

Short-answer questions

A short-answer question may test recall or it may test understanding by requiring you to undertake a short, one–stage calculation. Short-answer questions normally have space for the answers printed on the question paper. Here are some examples (the answers are shown in blue):

What is the relationship between electric current and charge flow?

Current = rate of flow of charge.

The current passing in a heater is 6 A when it operates from 240 V mains. Calculate the power of the heating element.

$P = I \times V = 6\,A \times 240\,V = 1440\,W$

Which of the following is the correct unit of acceleration?

A $m\,s^{-1}$
B $s\,m^{-1}$
C $m\,s^{-2}$
D $m^2\,s^{-1}$

C $m\,s^{-2}$

Structured questions

Structured questions are in several parts. The parts are usually about a common context and they often become progressively more difficult and more demanding as you work your way through the question. They may start with simple recall, then test understanding of a familiar or an unfamiliar situation. The most difficult part of a structured question is usually at the end, where the candidate is sometimes asked to suggest a reason for a particular phenomenon or social implication. Most of the practice questions in this book are structured questions, as this is the main type of question used in the assessment of A2 Level Physics.

When answering structured questions, do not feel that you have to complete one question before starting the next. The further you are into a question, the more difficult the marks are to obtain. If you run out of ideas, go on to the next question. Five minutes spent on the beginning of that question are likely to be much more fruitful than the same time spent wracking your brains trying to think of an explanation for an unfamiliar phenomenon.

Here is an example of a structured question that becomes progressively more demanding.

(a) A car speeds up from $20\,m\,s^{-1}$ to $50\,m\,s^{-1}$ in $15\,s$.

Calculate the acceleration of the car.

acceleration = increase in velocity ÷ time taken
= 30 m s⁻¹ ÷ 15 s = 2 m s⁻²

(b) The total mass of the car and contents is 950 kg.

Calculate the size of the unbalanced force required to cause this acceleration.

force = mass × acceleration
= 950 kg × 2 m s⁻² = 1900 N

(c) Suggest why the size of the driving force acting on the car needs to be greater than the answer to (b)

The driving force also has to do work to overcome the resistive forces, e.g. air resistance and rolling resistance.

Extended answers

In A2 Level Physics, questions requiring more extended answers will usually form part of structured questions. They will normally appear at the end of structured questions and be characterised by having at least three marks (and often more, typically five) allocated to the answers as well as several lines (up to ten) of answer space. These questions are also used to assess your abilities to communicate ideas and put together a logical argument.

The correct answers to extended questions are less well-defined than to those requiring short answers. Examiners may have a list of points for which credit is awarded up to the maximum for the question, or they may first of all judge the quality of your response as poor, satisfactory or good before allocating it a mark within a range that corresponds to that quality.

As an example of a question that requires an extended answer, a structured question on the use of solar energy could end with the following:

Suggest why very few buildings make use of solar energy in this country compared to countries in southern Europe. [5]

Points that the examiners might look for include:

• the energy from the Sun is unreliable due to cloud cover
• the intensity of the Sun's radiation is less in this country than in southern Europe due to the Earth's curvature
• more energy is absorbed by the atmosphere as the radiation has a greater depth of atmosphere to travel through
• fossil fuels are in abundant supply and relatively cheap
• the capital cost is high, giving a long payback time
• photo-voltaic cells have a low efficiency
• the energy is difficult to store for the times when it is needed the most

Full marks would be awarded for an argument that put forward three or four of these points in a clear and logical way.

Free-response questions

Little use is made of free-response and open-ended questions in A2 Level Physics. These types of question allow you to choose the context and to develop your own ideas. Examples could include 'Describe a laboratory method of determining *g*, the value of free-fall acceleration' and 'Outline the evidence that suggests that light has a wave-like behaviour'. When answering this type of question it is important to plan your response and present your answer in a logical order.

Exam technique

A2 Physics builds from AS Level Physics. This Study Guide has been written so that you will be able to tackle A2 Physics from an AS Physics background.

You should not need to search for important Physics from AS because this has been included where needed in each chapter. If you have not studied Physics for some time, you should still be able to learn Advanced Level Physics using this text and the AS Study Guide.

What are examiners looking for?

Examiners use instructions to help you to decide the length and depth of your answer.

If a question does not seem to make sense, you may have misread it – read it again!

State, define or list

This requires a short, concise answer, often recall of material that can be learnt by rote.

Explain, describe or discuss

Some reasoning or some reference to theory is required, depending on the context.

Outline

This implies a short response, almost a list of sentences or bullet points.

Predict or deduce

You are not expected to answer by recall but by making a connection between pieces of information.

Suggest

You are expected to apply your general knowledge to a 'novel' situation, one which you have not directly studied during the A2 Physics course.

Calculate

This is used when a numerical answer is required. You should always use units in quantities and significant figures should be used with care.

Look to see how many significant figures have been used for quantities in the question and give your answer to this degree of precision.

If the question uses 3 (sig figs), then give your answer to 3 (sig figs) also.

Some dos and don'ts

Dos

Do answer the question

- No credit can be given for good Physics that is irrelevant to the question.

Do use the mark allocation to guide how much you write

- Two marks are awarded for two valid points – writing more will rarely gain more credit and could mean wasted time or even contradicting earlier valid points.

Do use diagrams, equations and tables in your responses

- Even in 'essay-type' questions, these offer an excellent way of communicating physics.

Do write legibly

- An examiner cannot give marks if the answer cannot be read.

Do write using correct spelling and grammar. Structure longer essays carefully

- Marks are now awarded for the quality of your language in exams.

Don'ts

Don't fill up any blank space on a paper

- In structured questions, the number of dotted lines should guide the length of your answer.
- If you write too much, you waste time and may not finish the exam paper. You also risk contradicting yourself.

Don't write out the question again

- This wastes time. The marks are for the answer!

Don't contradict yourself

- The examiner cannot be expected to choose which answer is intended. You could lose a hard-earned mark.

Don't spend too much time on a part that you find difficult

- You may not have enough time to complete the exam. You can always return to a difficult calculation if you have time at the end of the exam.

What grade do you want?

Everyone would like to improve their grades but you will only manage this with a lot of hard work and determination. You should have a fair idea of your natural ability and likely grade in Physics and the hints below offer advice on improving that grade.

For a Grade A

You will need to be a very good all-rounder.

- You must go into every exam knowing the work extremely well.
- You must be able to apply your knowledge to new, unfamiliar situations.
- You need to have practised many, many exam questions so that you are ready for the type of question that will appear.

The exams test all areas of the syllabus and any weaknesses in your Physics will be found out. There must be no holes in your knowledge and understanding. For a Grade A, you must be competent in all areas.

For a Grade C

You must have a reasonable grasp of Physics but you may have weaknesses in several areas and you will be unsure of some of the reasons for the Physics.

- Many Grade C candidates are just as good at answering questions as the Grade A students but holes and weaknesses often show up in just some topics.
- To improve, you will need to master your weaknesses and you must prepare thoroughly for the exam. You must become a better all-rounder.

For a Grade E

You cannot afford to miss the easy marks. Even if you find Physics difficult to understand and would be happy with a Grade E, there are plenty of questions in which you can gain marks.

- You must memorise all definitions.
- You must practise exam questions to give yourself confidence that you do know some Physics. In exams, answer the parts of questions that you know first. You must not waste time on the difficult parts. You can always go back to these later.
- The areas of Physics that you find most difficult are going to be hard to score on in exams. Even in the difficult questions, there are still marks to be gained. Show your working in calculations because credit is given for a sound method. You can always gain some marks if you get part of the way towards the solution.

What marks do you need?

As a rough guide, you will need to score an average of 40% for a Grade E, 60% for a Grade C and 80% for a Grade A:

average	80%	70%	60%	50%	40%
grade	A	B	C	D	E

Essential mathematics

This section describes some of the mathematical techniques that are needed in studying A2 Level Physics.

Quantities and units

Physical quantities are described by the appropriate words or symbols, for example the symbol R is used as shorthand for the value of a *resistance*. The quantity that the word or symbol represents has both a numerical value and a unit e.g. 10.5 Ω. When writing data in a table or plotting a graph, only the numerical values are entered or plotted. For this reason headings used in tables and labels on graph axes are always written as (physical quantity)/(unit), where the slash represents division. When a physical quantity is divided by its unit, the result is the numerical value of the quantity.

Resistance is an example of a derived quantity and the ohm is a derived unit. This means that they are defined in terms of other quantities and units. All derived quantities and units can be expressed in terms of the seven base quantities and units that make up the SI, or International System of Units. The quantities, their units and symbols are shown in the table. The candela is not used in A Level Physics.

quantity	unit	symbol
length	metre	m
mass	kilogram	kg
time	second	s
electric current	ampere	A
temperature difference	kelvin	K
amount of substance	mole	mol
luminous intensity	candela	cd

Equations

The equations that you use in Physics are relationships between physical quantities. Since it is not possible for one quantity to equate to a different quantity, an equation must be homogenous, i.e. the units on each side of the equation must be the same. This is useful for:

- finding the units of a constant such as resistivity
- checking the possible correctness of an equation; if the units on each side are the same the equation may be correct but if they are different it is definitely wrong.

It is important to distinguish between a physical quantity and its unit. The value of a physical quantity includes both the numerical value and the unit it is measured in.

When working out the units of an equation, make it clear that you are dealing with units only by writing units of (quantity) = (unit).

Often equations need to be used in a different form to that in which they are given or memorised. The equation needs to be rearranged. When rearranging an equation it is important to remember that:

- the same mathematical operation must be applied to each side of the equation
- addition or subtraction should be done first, followed by multiplication or division and finally roots and powers.

For example, to rearrange the equation $v^2 = u^2 + 2as$ to enable u to be calculated:

1 subtract $2as$ from each side of the equation to give $v^2 - 2as = u^2$

2 square root each side of the equation to give $\sqrt{(v^2 - 2as)} = u$

An equation that relates the quantity to the product or quotient of two others can be rearranged using the 'magic triangle', a triangle divided into three.

To use the triangle, start with the quantities that are multiplied or divided in the relationship. For example, in $V = IR$, you would write I and R into the triangle first. Two quantities that are multiplied together are written side-by-side at the bottom of the triangle. Where one quantity is divided by the other these are written into the triangle as they would be in a division, with one quantity 'over the other'.

The remaining quantity is now written into the vacant slot. This is illustrated below.

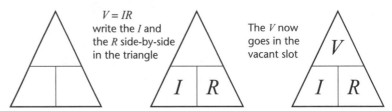

To use the triangle to make I the subject of the equation, cover up the I and you are left with 'V over R', so $I = V/R$.

Drawing graphs

Graphs have a number of uses in Physics:

- they give an immediate, visual display of the relationship between physical quantities
- they enable the values of quantities to be determined
- they can be used to confirm or disprove a hypothesis about the relationship between variables.

When plotting a graph, it is important to remember that values determined by experiment are not exact. They are subject to both the accuracy and precision of any measuring instrument and the person using it, as well as any changes that the measuring instrument itself might cause. Consider measuring the temperature of a liquid with a simple thermometer.

> Accuracy is the property of a measuring instrument. Precision refers to the reading taken; the precision depends on the user as well as the instrument being used.

- The manufacturer specifies the accuracy, the extent to which the thermometer is reliable. Typically, on a standard laboratory thermometer, this is ±1°C.
- When taking a reading, the user may read to the nearest half degree or nearest degree. This describes the precision.
- Inserting the thermometer into a substance may change the temperature of the substance as the two reach thermal equilibrium.

For these reasons, having plotted experimental values on a grid, the graph line is drawn as the best straight line or smooth curve that represents the points. Where there are 'anomalous' results, i.e. points that do not fit the straight line or curve, these should always be checked. If in doubt, ignore them but do add a note in your experimental work to explain why you have ignored them and suggest how any anomalous results could have arisen.

Measurements and graphs

Quantities such as velocity, acceleration and power are defined in terms of a **rate of change** of another quantity with time. This **rate of change** can be determined by calculating the gradient of an appropriate graph. For example, *velocity* is the *rate of change of displacement with time*. Its value is represented by the gradient of a displacement-time graph. Different techniques are used to determine the gradient of a straight line and a smooth curve. For a straight line:

A common error when determining the gradient of a graph is to work it out using the gridlines only, without reference to the scales on each axis.

- determine the value of Δy, the change in the value of the quantity plotted on the y-axis, using the whole of the straight line part of the graph
- determine the corresponding value of Δx
- calculate the gradient as $\Delta y \div \Delta x$

For a smooth curve, the gradient is calculated by first drawing a tangent to the curve and then using the above method to determine the gradient of the tangent. To draw a tangent to a curve:

- mark the point on the curve where the gradient is to be determined
- use a compass to mark in two points on the curve, close to and equidistant from the point where the gradient is to be determined
- join these points with a ruler and extend the line beyond each point.

These techniques are illustrated in the diagrams below.

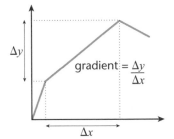
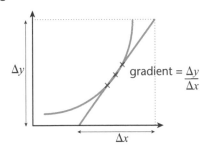

The area between a graph line and the horizontal axis, often referred to as the 'area under the graph' can also yield useful information. This is the case when the product of the quantities plotted on the axes represents another physical quantity, for example, on a *speed-time* graph this area represents the distance travelled. In the case of a straight line, the area can be calculated as that of the appropriate geometric figure. Where the graph line is curved, then the method of 'counting squares' is used.

- Count the number of complete squares between the graph line and the horizontal axis.
- Fractions of squares are counted as '1' if half the square or more is under the line, otherwise '0'.
- To work out the physical significance of each square, multiply together the quantities represented by one grid division on each axis.

These techniques are illustrated in the diagrams below

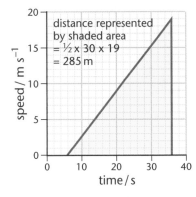
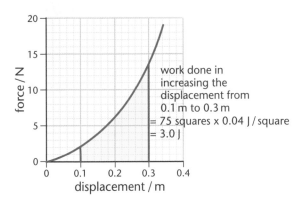

Equations from graphs

By plotting the values of two variable quantities on a suitable graph, it may be possible to determine the relationship between the variables. This is straightforward when the graph is a straight line, since all straight line graphs have an equation of the form $y = mx + c$, where m is the gradient of the graph and c is the value of y when x is zero, i.e. the intercept on the y-axis. The relationship between the variables is determined by finding the values of m and c.

The straight line graph below shows how the displacement, *s*, of an object varies with time, *t*.

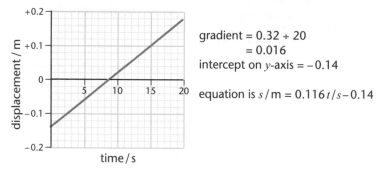

gradient = 0.32 ÷ 20
= 0.016
intercept on *y*-axis = −0.14

equation is $s/\text{m} = 0.116\,t/s - 0.14$

If the relationship between two variables is known to be of the form $y = kx^n$, the values of *k* and *n* can be determined by drawing a logarithmic graph. Since log *y* = log *k* + *n*log *x*, a graph of log *y* against log *x* has gradient equal to *n* and the intercept on the *y* axis is equal to log *k*.

In the case of an exponential change such as in capacitor discharge or radioactive decay, where a relationship is of the form $y = e^{-kx}$, the natural log function, ln, should be used. Applying this function to each side of the equation gives (since natural log and exponential are inverse functions) ln *y* = −*kx*. The gradient of a plot of ln *y* against *x* represents the constant *k*.

Trigonometry and Pythagoras

In the right-angled triangle shown below, the sides are labelled o (opposite), a (adjacent) and h (hypotenuse). The relationships between the size of the angle θ and the lengths of these sides are:

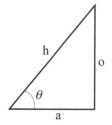

- $\sin\theta = o/h$
- $\cos\theta = a/h$
- $\tan\theta = o/a$

Pythagoras' theorem gives the relationship between the sides of a right-angled triangle: $h^2 = a^2 + o^2$

Multiples and submultiples

For A2 Physics, you are expected to be familiar with the following multiples of units:

name	multiple	symbol
micro-	10^{-6}	mμ
milli-	10^{-3}	m
kilo-	10^{3}	k
mega-	10^{6}	M
giga-	10^{9}	G

For example, the symbol MHz means 1×10^6 Hz and mN means 1×10^{-3} N.

Equations that you need to know

This is a list of equations that may be needed but are not provided in topic tests or end-of-course examinations. Not all of these equations are needed for every specification as some are in the AS not the A2 specification. The table shows the equations needed for each A2 specification. Where an equation is shown as being required in the AS specification, it could be required in the synoptic assessment at A2.

All other equations that are needed will be provided on a data sheet, along with any data such as the values of physical constants.

equation	symbol form	AQA A	AQA B	ED A	ED B	OCR A	OCR B	WJEC	NICCEA
speed = $\dfrac{\text{distance}}{\text{time taken}}$	$v = \dfrac{\Delta d}{t}$	AS	AS	AS	AS	AS	AS	AS	AS
force = mass × acceleration	$F = ma$	AS	AS	AS	AS	AS	AS	AS	AS
acceleration = $\dfrac{\text{change in velocity}}{\text{time taken}}$	$a = \dfrac{\Delta v}{t}$	AS	AS	AS	AS	AS	AS	AS	AS
density = $\dfrac{\text{mass}}{\text{volume}}$	$\rho = \dfrac{m}{V}$	AS		AS	AS	•			
momentum = mass × velocity	$p = mv$	AS	•	AS	•	•	•	•	AS
work done = force × distance moved in direction of force	$W = Fs$	AS	•	AS	AS	AS	AS	A2	•
power = $\dfrac{\text{energy transferred}}{\text{time taken}} = \dfrac{\text{work done}}{\text{time taken}}$	$P = \dfrac{\Delta W}{t}$	AS	•	AS	AS	AS	•	•	•
weight = mass × gravitational field strength	$W = mg$	•	•	AS	AS	AS	•	•	•
kinetic energy = ½ × mass × speed²	$E_k = \tfrac{1}{2}mv^2$	AS	AS	AS	AS	•	AS	AS	AS
change in gravitational potential energy = mass × gravitational field strength × change in height	$E_p = mgh$	AS	AS	AS	AS	•	•	AS	AS
pressure = $\dfrac{\text{force}}{\text{area}}$	$P = \dfrac{F}{A}$			AS		AS	AS		
pressure × volume = number of moles × molar gas constant × absolute temperature	$pV = nRT$	AS	•	AS	•	•	•	•	•
charge = current × time	$q = It$	AS	AS	AS	AS	AS	AS	AS	AS
potential difference = current × resistance	$V = IR$	AS	AS	AS	AS	AS	AS	AS	AS
electrical power = potential difference × current	$P = IV$	AS	AS	AS	AS	AS	AS	AS	AS
potential difference = $\dfrac{\text{energy transferred}}{\text{charge}}$	$V = \dfrac{W}{q}$	AS	AS	AS	AS	AS	AS	AS	AS
resistance = $\dfrac{\text{resistivity} \times \text{length}}{\text{cross-sectional area}}$	$R = \dfrac{\rho l}{A}$	AS	AS	AS	AS	AS	AS	AS	AS
energy = potential difference × current × time	$E = VIt$	AS	AS	AS	AS	AS	AS	AS	AS
wave speed = frequency × wavelength	$v = f\lambda$	•	AS	•	AS	AS	AS	AS	AS
centripetal force = $\dfrac{\text{mass} \times \text{speed}^2}{\text{radius}}$	$F = \dfrac{mv^2}{r}$	•	•	•	•	•	•	•	•
force between two charges	$F = \dfrac{kQ_1 Q_2}{r^2}$	•	•	•	•	•	•	•	•
force between two masses	$F = \dfrac{Gm_1 m_2}{r^2}$	•	•	•	•	•	•	•	•
capacitance = $\dfrac{\text{charge stored}}{\text{voltage}}$	$C = \dfrac{Q}{V}$	•	•	•	•	•	•	•	•
the transformer equation	$\dfrac{V_p}{V_s} = \dfrac{N_p}{N_s}$		•	•	•		•		•

Four steps to successful revision

Step 1: Understand

- Study the topic to be learned slowly. Make sure you understand the logic or important concepts.
- Mark up the text if necessary – underline, highlight and make notes.
- Re-read each paragraph slowly.

GO TO STEP 2

Step 2: Summarise

- Now make your own revision note summary:
  What is the main idea, theme or concept to be learned?
  What are the main points? How does the logic develop?
  Ask questions: Why? How? What next?
- Use bullet points, mind maps, patterned notes.
- Link ideas with mnemonics, mind maps, crazy stories.
- Note the title and date of the revision notes
  (e.g. Physics: Mechanics, 3rd March).
- Organise your notes carefully and keep them in a file.

This is now in **short term memory**. You will forget 80% of it if you do not go to Step 3.
GO TO STEP 3, but first take a 10 minute break.

Step 3: Memorise

- Take 25 minute learning 'bites' with 5 minute breaks.
- After each 5 minute break test yourself:
  Cover the original revision note summary
  Write down the main points
  Speak out loud (record on tape)
  Tell someone else
  Repeat many times.

The material is well on its way to **long term memory**.
You will forget 40% if you do not do step 4. **GO TO STEP 4**

Step 4: Track/Review

- Create a Revision Diary (one A4 page per day)
- Make a revision plan for the topic, e.g. 1 day later, 1 week later, 1 month later.
- Record your revision in your Revision Diary, e.g.
  Physics: Mechanics, 3rd March 25 minutes
  Physics: Mechanics, 5th March 15 minutes
  Physics: Mechanics, 3rd April 15 minutes
  ... and then at monthly intervals.

Chapter 1

Mechanics

The following topics are covered in this chapter:

- *Newton I and III*
- *Momentum and the second law*
- *Energy to work*
- *Motion in a circle*

1.1 Newton I and III

After studying this section you should be able to:

- *state Newton's first and third laws of motion*
- *apply the first law to situations where the forces acting on an object are balanced*
- *apply the third law to the forces acting on two objects that interact*

LEARNING SUMMARY

Newton's laws of motion

With his three laws of motion, Newton aimed to describe and predict the movement of every object in the Universe. Like many other 'laws' in physics, they give the right answers much of the time. The paths of planets, moons and satellites are all worked out using Newton's laws.

This section looks at formal statements of the first and third law and how they apply to everyday situations.

The first law

AQA A	AS	EDEXCEL B	AS
AQA B	AS	OCR A	M4
EDEXCEL A	AS	WJEC	M4

The phrase 'uniform motion' means moving in a straight line at constant speed, i.e. moving at constant velocity.

KEY POINT

Newton's first law states that:

An object maintains its state of rest or uniform motion unless there is a resultant, or unbalanced, force acting on it.

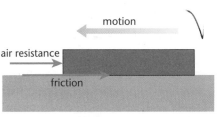

friction and air resistance create a resultant force on this sliding object

At first, this seems contrary to everyday experience. Give an object a push and it slows down before coming to rest. This is not 'continuing in a state of uniform motion'. Newton realised that in this case the unseen resistive forces of friction and air resistance together act in opposition to the motion; as there is no longer a driving force after the object has been pushed, there is a resultant backwards-directed force acting on it.

Motion without resistive forces is difficult to achieve on Earth: there is always air resistance or friction from a surface that a moving object rests on. The nearest that we can get to modelling motion with no resistive forces is to study motion on ice or an air track.

The converse of Newton's first law is also true:

if an object is at rest or moving at constant velocity the resultant force on it is zero

Key point from AS

- When the forces on an object are balanced, the vector diagram is a closed figure.
Revise AS section 1.1

This means that where there are two forces acting on an object that satisfies these conditions, they must be equal in size and opposite in direction. If there are three or more forces acting then the vector diagram is a closed figure.

One example of the first law is a vehicle moving at constant velocity; the **driving force** and **resistive force** are equal in size and act in opposite directions. The diagram shows an example of a situation where three forces sum to zero; it illustrates a vehicle parked on a hill at rest and a vector triangle that shows the forces.

A normal reaction force acts on all four wheels; the arrow shown on the diagram represents the sum of these forces.

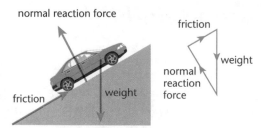

The third law

AQA A	AS	EDEXCEL B	AS
AQA B	AS	OCR A	M4
EDEXCEL A	AS	WJEC	M4

The laws were originally written in Latin. The third law is directly translated as 'to every action there is an equal and opposite reaction'. This was widely misinterpreted as meaning that two equal and opposite forces act on the same object, resulting in zero acceleration.

This is both the simplest in its statement and the most misunderstood of the three laws. The statement given here is not a direct translation of the original, but it helps to remove some of the misunderstanding.

> **KEY POINT**
>
> Newton's third law can be stated as:
>
> If object A exerts a force on object B, then B exerts a force equal in size and opposite in direction on A.

According to the third law, forces do not exist individually but in pairs. However, it is important to remember that:

- the forces are of the same type, i.e. both gravitational or electrical
- the forces act on different objects
- the third law applies to all situations.

Some examples

AQA A	AS	EDEXCEL B	AS
AQA B	AS	OCR A	M4
EDEXCEL A	AS	WJEC	M4

Application of Newton's third law can lead to some surprising results. If you step off a wall the Earth pulls you down towards the ground. According to the third law, you also pull the Earth up with an equal-sized force. So why doesn't the Earth accelerate upwards to meet you instead of the other way round? The answer is it does, but if you apply $F = ma$ to work out the Earth's acceleration, it turns out to be minimal.

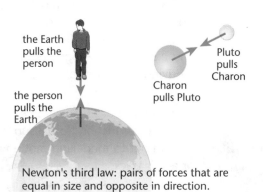

Newton's third law: pairs of forces that are equal in size and opposite in direction.

This type of situation is where the greatest misunderstanding of Newton's third law occurs. Remember, third law pairs of forces act on **different** objects.

Planets pull moons and, according to the third law, moons pull planets with an equal-sized force. Why does the moon go round the planet instead of the other way round?

Again, the answer is it does. In fact, they both rotate around a common centre of mass. In the case of the Earth and its Moon, the centre of mass is so close to the centre of the Earth that to all intents and purposes the Moon orbits the Earth.

The diagram shows another pair of forces that are equal in size and opposite in direction. This is an application of the first law – the vase is in a state of rest so there is no resultant force acting on it.

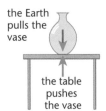

the Earth pulls the vase

the table pushes the vase

These forces are also equal in size and opposite in direction – an application of Newton's first law, not third.

Progress check

1 Explain why a driving force is needed to maintain motion on the surface of the Earth. Because of the opposite force of friction

2 Suggest why the driving force needed to maintain motion on the surface of the Moon is less than for the same vehicle on the surface of the Earth. because there is a less force caused by gravity on the moon no

3 A foot kicks a ball. What is the force that makes up the pair in the sense of the third law?

3 The ball pushes the foot.

2 There is no air resistance on the Moon.

1 The driving force is needed to balance the resistive forces that oppose motion.

1.2 Momentum and the second law

After studying this section you should be able to

- *apply the principle of conservation of momentum to collisions in one dimension*
- *state the relationship between the change of momentum of an object and the resultant force on it*
- *explain the difference between an elastic and an inelastic collision*

Momentum

AQA A	AS	OCR A	M4
AQA B	M4	OCR B	M4
EDEXCEL A	AS	NICCEA	M4
EDEXCEL B	M4	WJEC	M4

A collision between two objects need not involve physical contact. Imagine two positive charges approaching each other; the repulsive forces could cause them to reverse in direction without actually touching.

Which is more effective at demolishing a brick wall, a 1 kg iron ball moving at 50 m s^{-1} or a 1000 kg iron ball moving at 1 m s^{-1}? Both the mass and the velocity need to be taken into account to answer this question.

Newton realised that what happens to a moving object involved in a collision depends on two factors:

- the mass of the object
- the velocity of the object.

He used the concept of **momentum** to explain the results of collisions between objects.

> **momentum = mass × velocity**
>
> $$p = m \times v$$
>
> Momentum is a vector quantity, the direction being the same as that of the velocity. It is measured in N s or kg m s^{-1}; these units are equivalent.

A ten-pin bowling ball has a mass of several kilogrammes, so even at low speeds it has much more momentum that a tennis ball (mass 0.06 kg) travelling as fast as you can throw it.

Everything that moves has momentum and exerts a force on anything that it interacts with. This applies as much to the light that is hitting you at the moment as it does to a collision between two vehicles.

No matter how hard you throw a tennis ball at a garden wall you will not knock the wall down; nor will you be very successful if you use a tennis ball to play ten-pin bowling!

Conservation of momentum

AQA A	AS	OCR A	M4
AQA B	M4	OCR B	M4
EDEXCEL A	AS	NICCEA	M4
EDEXCEL B	M4	WJEC	M4

The phrase 'equal and opposite' is used as a shorthand way of writing 'equal in size and opposite in direction'.

When a force causes an object to change its velocity, there is also a change in momentum. When two objects interact or collide, they exert equal and opposite forces on each other and so the momentum of each one changes. This is illustrated in the diagram, where the blue ball loses momentum and the black ball gains momentum.

before

5.0 m s^{-1} 3.0 m s^{-1}

0.4 kg   0.1 kg

after

4.2 m s^{-1} 6.2 m s^{-1}

Like the forces they exert on each other during the collision, the changes in momentum of the balls are *equal in size and opposite in direction*.

It follows that:

- the combined momentum of the balls is the same before and after they collide.

To satisfy the conditions for conservation of momentum to apply, the only forces acting on the objects must be the ones between the objects themselves.

This is why conservation of momentum is usually demonstrated on an air track or other low friction surface.

This is an example of the principle of conservation of momentum.

> **KEY POINT**
>
> The principle of conservation of momentum states that:
>
> When two or more objects interact the total momentum remains constant provided that there is no external resultant force.

In the context of the colliding balls, an external resultant force could be due to friction or something hitting one of the balls. Either of these would result in a change in the total momentum.

Different types of interaction

AQA A	AS	OCR A	M4
AQA B	M4	NICCEA	M4
EDEXCEL A	AS	WJEC	M4
EDEXCEL B	M4		

Objects moving in opposite directions

When two objects approach each other and 'collide' head-on, the result of the collision depends on whether they stick together or whether one or both rebound.

The diagram shows two 'vehicles' approaching each other on an air track. The vehicles join together and move as one.

To make two air track vehicles stick together, one has some Plasticine stuck to it and the other is fitted with a pin that sticks into the Plasticine.

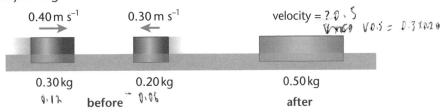

The principle of conservation of momentum can be used to work out the combined velocity, v, after the collision.

Taking velocity from left to right as being positive:

total momentum before collision =
$$0.3 \text{ kg} \times + 0.4 \text{ m s}^{-1} + 0.2 \text{ kg} \times - 0.3 \text{ m s}^{-1} = 0.06 \text{ N s}.$$

This must equal the momentum after the collision, so $0.5 \text{ kg} \times v = 0.06 \text{ N s}$,

i.e. $v = 0.12 \text{ m s}^{-1}$. As this is positive, it follows that the 'vehicle' is moving from left to right.

Rebound

When a light object collides with a heavier one, the light object may rebound, reversing the direction of its momentum. The diagram shows an example of such a collision.

A rebound collision can be modelled on an air track by fitting the vehicles with repelling magnets.

The figures here are rounded. The 3.33 is a rounded value of 3⅓.

Application of the principle of conservation of momentum in this case gives:

$$2 \text{ kg} \times 10 \text{ m s}^{-1} + 10 \text{ kg} \times 0 \text{ m s}^{-1} = 2 \text{ kg} \times v + 10 \text{ kg} \times 3.33 \text{ m s}^{-1}$$

So $v = - 6.7 \text{ m s}^{-1}$. The significance of the negative sign is that the green ball is now travelling from right to left.

When one object rebounds off another during a collision, both objects undergo a greater change in momentum than if they stick together. In the example above, the size of the change in momentum of the small ball is $2 \text{ kg} \times 10 \text{ m s}^{-1} - 2 \text{ kg} \times - 6.7 \text{ m s}^{-1} = 33.3 \text{ m s}^{-1}$. Had it carried on moving from left to right, the change in momentum would have been much smaller than this.

Recoil

When fire fighters use water to put out a fire, it often takes two people to hold the hose. This is due to the recoil as water is forced out of the nozzle at high speed.

In some situations two objects are stationary before they interact, having a combined momentum of zero. After they interact, both objects move off but the combined momentum is still zero. Examples of this include:

- a stationary nucleus undergoes radioactive decay, giving off an alpha particle
- a person steps off a boat onto the quayside
- two ice skaters stand facing each other; then one pushes the other
- two air track 'vehicles' fitted with repelling magnets are held together and then released.

In each case, both objects start moving from rest. For the combined momentum to remain zero, each object must gain the same amount of momentum but these momentum gains must be in *opposite directions*.

In recoil situations, each object gains the same amount of momentum, but in opposite directions. It follows that the one with the smaller mass ends up with the faster speed.

The ice skater who does the pushing **recoils** as an equal and opposite force is exerted by the one who is pushed.

In the example shown in the diagram above, an 80 kg man pushes a 50 kg child. Their combined momentum is zero both before and after this interaction.

Taking velocity from left to right as positive, this means that:

$$80 \text{ kg} \times -0.6 \text{ m s}^{-1} + 50 \text{ kg} \times v = 0$$

So v, the velocity of the child after the interaction, is 0.96 m s^{-1} to the right.

Elastic or inelastic

AQA A	AS	NICCEA	M4
EDEXCEL B	M4	WJEC	M4

It is important to emphasise here that total energy is **always** conserved.

In all three examples above, both momentum and energy are conserved. This is always the case. A collision where the total **kinetic** energy before and after the interaction is the same is called an **elastic** collision. Where there is a gain or loss of kinetic energy as a result of the interaction the collision is called **inelastic**.

In an elastic collision: momentum, kinetic energy and total energy are conserved.

In an inelastic collision: momentum and total energy are conserved, but the amount of kinetic energy changes.

Of the examples given, only the second is an elastic collision.

Momentum and the second law

AQA A	AS	OCR A	M4
AQA B	M4	OCR B	M4
EDEXCEL A	AS	NICCEA	M4
EDEXCEL B	M4	WJEC	M4

The expression $\Delta p/\Delta t$ can be read as: change in momentum ÷ time taken for the change.

Newton's second law establishes the relationship between the resultant force acting on an object and the change of momentum that it causes. An important result that follows from the second law and the definition of the newton is $F = m \times a$.

> **KEY POINT**
>
> Newton's second law states that:
>
> The rate of change of momentum of an object is proportional to the resultant force acting on it and acts in the direction of the resultant force.
>
> $$\Delta p/\Delta t \propto F$$

Key points from AS

- **The definition of the size of the newton.**
  Revise AS section 1.6

The definition of the size of the newton fixes the proportionality constant at one. This enables the second law to be written as:

> **KEY POINT**
>
> Force = rate of change of momentum
> $F = \Delta p / \Delta t$

In this form, the second law is useful for working out the force in situations such as jet and rocket propulsion where the change in momentum each second can be calculated easily.

A rocket carries its own oxygen supply, so that it can fire the engines when it is travelling in a vacuum. A spacecraft travelling where it is not affected by any gravitational fields or resistive forces maintains a constant velocity, so it only needs to fire the rocket engines to change speed or direction.

Jet and rocket engines use the same principle; hot gases are ejected from the back of the engine to provide a force in the forwards direction. The difference is that jet engines are designed to operate within the Earth's atmosphere, so they can take oxygen from their surroundings.

The principle of rocket propulsion can be seen by blowing up a balloon and letting it go; the air squashed out of the neck of the balloon gains momentum as it leaves, causing the balloon and the air that remains inside to gain momentum at the same rate but in the opposite direction. The diagram shows the principle of rocket propulsion.

The unit kg m s^{-2} is equivalent to the N, so a rate of change of momentum of 4 500 000 kg m s^{-2} is another way of stating: a force of 4 500 000 N.

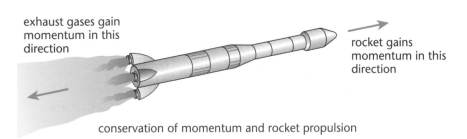

exhaust gases gain momentum in this direction

rocket gains momentum in this direction

conservation of momentum and rocket propulsion

In this example, the exhaust gases are emitted at a rate of 3000 kg s^{-1}. The speed of the exhaust gases relative to the rocket is 1500 m s^{-1}. The rate of change of momentum of these gases is therefore 4 500 000 kg m s^{-2}, i.e. 4 500 000 kg m s^{-1} each second. This is the size of the force on both the gases and the rocket.

Progress check

1   Two vehicles fitted with repelling magnets are held together on an air track. The masses of the vehicles are 0.10 kg and 0.15 kg.

   The vehicles are released. The 0.10 kg vehicle moves to the left with a speed of 0.24 m s^{-1}. Calculate the velocity of the heavier vehicle.

2   A ball of mass 0.020 kg hits a wall at a speed of 12.5 m s^{-1} and rebounds at a speed of 10.0 m s^{-1}.
   a   Calculate the change in momentum of the ball.
   b   If the ball is in contact with the wall for 0.010 s, calculate the size of the force between the ball and the wall.
   c   Explain whether this collision is elastic or inelastic.

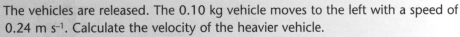

c   Inelastic. The ball has less kinetic energy after rebound than before it hits the wall.
   b   45 N
2 a   0.45 N s in the direction of the rebound
1   0.16 m s^{-1} to the right

1.3 Energy to work

After studying this section you should be able to:

- *explain the energy transfer that takes place when a force causes motion*
- *calculate kinetic energy and changes in gravitational potential energy*
- *apply the principle of conservation of energy*

LEARNING SUMMARY

Working

AQA A	AS	OCR A	M4
AQA B	M4	OCR B	M4
EDEXCEL A	AS	NICCEA	M4
EDEXCEL B	AS	WJEC	M4

Every event requires **work** to make it happen. Even if you relax and try to do nothing, your body is working just to keep you alive. Any force that causes movement is doing work. Pushing a supermarket trolley is working, as is throwing or kicking a ball. However, holding some weights above your head is not working; it may cause your arms to ache, but the force on the weights is not causing any movement!

How much work a force does depends on:

- the size of the force
- the direction of movement
- the distance that an object moves.

The phrase 'distance moved in the direction of the force' is used here instead of 'displacement' to emphasise that the force and displacement it causes are measured in the same direction.

> **KEY POINT**
>
> The work done by a force that causes movement is defined as:
>
> work = force × distance moved in the direction of the force
>
> $$W = F \times s$$
>
> Work is measured in joules (J) where 1 joule is the work done when a force of 1 N moves its point of application 1 m in the direction of the force.

Here are some examples of forces that are working and forces that are not working:

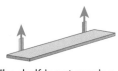

The tension in the string is not working as there is no movement **in the direction of the force.**

The shelf is not moving, so no work is being done.

A pylon that supports electricity transmission cables is not working, but a wind that causes the cables to move is!

This force causes movement **in the direction of the force.**

The horizontal component of the force on the log is doing the work here.

In circular motion, the direction of the velocity is along a tangent to the circle.

In the case of the stone being whirled in a horizontal circle, the stone's velocity is always at **right angles** to the force that maintains the motion, so the centripetal force is not working.

Although the force pulling the log and the movement it causes are not in the same direction, the force on the log has a component in the same direction as the log's movement. In calculating the work done the horizontal component of the force is multiplied by the horizontal distance moved by the log.

Work and energy transfer

AQA A	AS	OCR A	M4
AQA B	M4	NICCEA	M4
EDEXCEL A	AS	WJEC	M4
EDEXCEL B	AS		

The physical processes of **working** and **energy transfer** are inseparable. Work requires an energy source, for example a fuel, an electric cell or a wound-up spring. As the work is done, energy leaves the source and ends up elsewhere, this is what is meant by **energy transfer**.

> An object has energy if it can exert a force that causes movement of the point of application of the force.

> **KEY POINT**
>
> Energy is the ability to do work.
> The energy transfer from a source is equal to the amount of work done.
> Like work, energy is measured in joules.

Here are some examples of everyday energy transfers.

A bus accelerates away from a bus stop and then maintains a steady speed

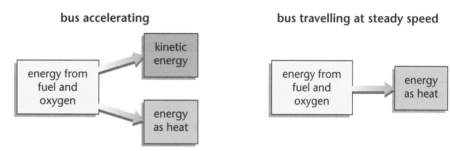

> A common error is to describe the energy transfer of a vehicle moving at a constant speed as energy from the fuel and oxygen being transferred to kinetic energy.

As the bus speeds up, some of the energy from the fuel and oxygen is transferred to **kinetic energy** of the bus, the energy it has due to its movement. The rest of the energy is transferred to the surroundings. This takes place in a number of ways; the exhaust gases transfer energy as heat directly into the surrounding air and some energy is also transferred to heat wherever there are resistive forces.

Once the bus is maintaining a steady speed there is no increase in its kinetic energy, so all the energy transferred from the fuel and oxygen is going to the surroundings.

An electric motor lifts a load at a steady speed

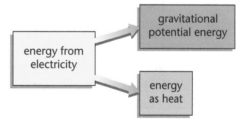

> The phrase 'gravitational potential energy' is often abbreviated to 'gpe' or just 'potential energy', although strictly speaking 'potential energy' can be any form of stored energy, for example in a spring or a fuel.

The load gains **gravitational potential energy** as it is lifted, the energy it has due to its position above the Earth's surface. All the movement is at a constant speed, so there is no change in the kinetic energy of the system. As with the bus travelling at a steady speed, work done against resistive forces causes heating, and this energy is transferred as heat to the surroundings.

Energy and a filament lamp

When a filament lamp is switched on, some energy is absorbed by the filament as it reaches its operating temperature. This takes a fraction of a second. Some energy is absorbed by the glass envelope and the gas inside. These take a little longer to reach a steady temperature, perhaps a few seconds.

Once the lamp has reached its steady state, of the 60 J of energy passing into the lamp each second from the electricity supply, typically 3 J passes out as light. This

In the steady state, the parts of the lamp are emitting energy and absorbing energy at the same rate, so there is no change in temperature.

Sankey diagrams are a useful way of showing the energy flow through a process where there are several stages, for example the energy flow through a power station.

is shown in the diagram. The Sankey diagram gives a visual indication of the relative proportions of energy transferred into a desirable output and wasted.

Energy flow through filament lamp

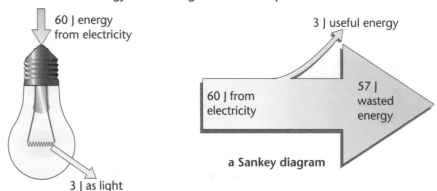

a Sankey diagram

The 'wasted' energy is transferred to the surroundings in two main ways:

* from the glass envelope to the surrounding air by conduction and convection
* from all parts of the lamp as non-visible electromagnetic radiation (mainly infra-red).

The light and the infra-red radiation all cause heating when they are absorbed, so all the energy input to the lamp ends up as heat!

These examples illustrate some important points about energy transfer:

* energy due to movement is called kinetic energy
* energy due to position above the Earth's surface is called gravitational potential energy, often abbreviated to potential energy
* work usually involves some transfer of energy to the surroundings.

Conservation of energy

AQA A	AS	OCR A	M4
AQA B	M4	OCR B	M4
EDEXCEL A	AS	NICCEA	M4
EDEXCEL B	AS	WJEC	M4

The examples of energy transfer identify some ways in which energy flows into and out of a process where work is done. Like mass and charge, energy is a conserved quantity.

Take care not to confuse 'conservation of energy' with 'conservation of energy resources'. Energy resources such as coal can be used up, and conservation in this context means preserving them as long as possible.

> **KEY POINT**
>
> The principle of **conservation of energy** states that:
> energy can not be created or destroyed.

This simple statement means that energy is never **used up**, although often the energy output from a working process is in a form which is difficult to harness to do more work. The energy output from a filament lamp, for example, is effectively **lost** as heat in the surroundings. Like the energy from the bus, it raises the temperature of its surroundings by such a small amount that makes it difficult to recover.

In many energy transfer processes the amount of energy at each stage cannot easily be quantified; it is difficult to put a figure on the total kinetic and potential energy of the moving parts in an engine for example. There are simple formulas for calculating the kinetic energy and gravitational potential energy of individual objects.

The formula for change in gravitational potential energy is only valid for changes in height close to the Earth's surface, where g has a constant value.

> **KEY POINT**
>
> kinetic energy = ½ × mass × (speed)²
> $$E_k = \tfrac{1}{2} mv^2$$
> change in gravitational potential energy = weight × change in height
> $$\Delta E_p = mg\Delta h$$

The formula for calculating gravitational potential energy gives the change in energy because 'ground level' varies and so does not provide an accurate reference point for zero energy. This is not the case with kinetic energy as 'no movement relative to the Earth's surface' is the same the world over.

When a streamlined object is falling freely in the atmosphere at a low speed, the resistive forces are small and so little energy is transferred to the surroundings. In this case the principle of conservation of energy can be applied to the transfer of gravitational potential energy to kinetic energy. When a parachutist is falling at a constant speed conservation of energy still applies, but the energy transfer is from gravitational potential energy to heat in the atmosphere.

The principle of conservation of energy applies to all energy transfers.

Energy transfer of falling objects

A ball falling freely towards the Earth

gravitational potential energy ➜ kinetic energy

A parachutist falling at constant speed

gravitational potential energy ➜ heat in the surrounding air

Progress check

A 2.0 kg mass is released from a height of 8.0 m above the Earth's surface and allowed to fall to the ground. Assuming $g = 10$ m s^{-2}, calculate:

a the loss of gravitational potential energy
b the gain in kinetic energy
c the speed of the mass as it hits the ground.

a 160 J
b 160 J
c 12.6 m s^{-1}

1.4 Motion in a circle

After studying this section you should be able to:

- use the radian to measure angular displacement
- describe the forces on an object moving in a circle
- recall and use the expressions for centripetal force and centripetal acceleration

LEARNING SUMMARY

Measuring angles

AQA A	M4	OCR A	M4
AQA B	M4	NICCEA	M4
EDEXCEL A	M4	WJEC	M4
EDEXCEL B	M4		

When an object travels through a complete circle it moves through an angle of 360°, although the distance it travels depends on the radius of the circle. When a train or motor vehicle travels round a bend, the outer wheels have further to travel than the inner wheels. Similarly, when a compact disc is being played all points on the disc move through the same angle in any given time, but the linear speed at any point depends on its distance from the centre.

An alternative unit to the degree for measuring angular displacement is the radian, defined as the arc length divided by the radius of the circle. This is shown in the diagram.

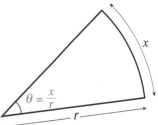

$$\theta = \frac{x}{r}$$

Measuring an angle in radians

> To convert an angle in degrees to one in radians, multiply by π and divide by 180.

It follows that movement through one complete circle, or 360°, is equal to a displacement of θ = circumference ÷ radius = $2\pi r \div r = 2\pi$ radians.

Speed in a circle

AQA A	M4	OCR A	M4
AQA B	M4	OCR B	M4
EDEXCEL A	M4	NICCEA	M4
EDEXCEL B	M4	WJEC	M4

When an object moves in a circular path its velocity is always at right angles to a radial line, a line joining it to the centre of the circle.

Two objects, such as points on a CD, at different distances from the centre each complete the same number of revolutions in any specified time but they have different speeds. They have the same angular velocity but different linear velocities.

> Angular velocity is a vector quantity that can have one of two directions: clockwise or anticlockwise.

KEY POINT

Angular velocity, ω, is defined as:

the rate of change of angular displacement with time.

It is represented by the gradient of an angular displacement–time graph.

average angular velocity = angular displacement ÷ time
$$\omega = \Delta\theta/\Delta t$$
where $\Delta\theta$ is the angular displacement in time Δt.

The relationship between angular velocity and linear velocity for an object moving in a circle of radius r is:

$$v = r\omega.$$

The time period for one revolution, T, is equal to the angular displacement (2π radians) divided by the angular velocity:

$$T = 2\pi/\omega.$$

The frequency of a circular motion, f, is equal to $1/T$:
$$f = 1/T = \omega/2\pi.$$

Centripetal acceleration

AQA A	M4	OCR A	M4
AQA B	M4	OCR B	M4
EDEXCEL A	M4	NICCEA	M4
EDEXCEL B	M4	WJEC	M4

Any object that changes its velocity is accelerating. Since velocity involves both speed and direction, a change in either of these quantities is an acceleration. An object moving at a constant speed in a circle is therefore accelerating as it is continually changing its direction. The vector diagram below shows the change in velocity, Δv, for an object following a circular path. This change in direction is always towards the centre of the circle.

> Note that Δv is the change in velocity from v_1 to v_2, so
> $v_1 + \Delta v = v_2$

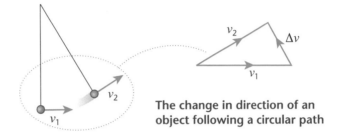

The change in direction of an object following a circular path

The acceleration of an object moving in a circle:

- is called **centripetal acceleration**
- is directed towards the centre of the circle.

> **KEY POINT**
>
> Centripetal acceleration, a, is the acceleration of an object moving in a circular path.
> $$a = v^2/r = r\omega^2$$
> where v is the linear velocity, ω is the angular velocity and r is the radius of the circle.

The unbalanced force

AQA A	M4	OCR A	M4
AQA B	M4	OCR B	M4
EDEXCEL A	M4	NICCEA	M4
EDEXCEL B	M4	WJEC	M4

To make an object accelerate, there has to be an unbalanced force acting in the same direction as the acceleration. In the case of circular motion the unbalanced force is called the **centripetal force**.

> It is important to understand that the centripetal force is the resultant of all the forces acting on an object. A common misconception is that it is an extra force that appears when an object moves in a circular path.

When a train travels around a bend, the centripetal force comes from the push of the outer rail on the wheel. For a road vehicle such as a car, bus or bicycle it is the push of the road on the tyres; if there is insufficient friction due to ice or a slippery surface, the vehicle does not complete the turn and may leave the road.

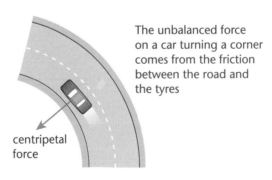

The unbalanced force on a car turning a corner comes from the friction between the road and the tyres

centripetal force

The size of the unbalanced force required to maintain circular motion can be calculated using $F = ma$:

> **KEY POINT**
>
> An object moving at constant speed in a circular path requires an unbalanced force towards the centre of the circle.
> $$\text{centripetal force, } F = mv^2/r = mr\omega^2$$

The theme park ride

AQA A	M4	OCR A	M4
AQA B	M4	OCR B	M4
EDEXCEL A	M4	NICCEA	M4
EDEXCEL B	M4	WJEC	M4

You feel the apparent change in weight when you travel in a lift. As the lift accelerates upwards you feel heavier because the normal contact force is greater than your weight.

Key points from AS

• **Vector diagrams**
  Revise AS section 1.1

Check that in each case the resultant of the forces is equal to mv^2/r towards the centre of the circle.

Another possibility is that the speed of the ride is such that the required centripetal force is less than the person's weight. In this case a safety harness is required to stop the person from falling out.

Some theme park rides involve motion in a vertical circle. Part of the thrill of being on one of these rides is the apparent change in weight as you travel round the circle. This is because the force that you feel as you sit or stand is not your weight, but the normal contact force pushing up. If the forces on you are balanced, this force is equal in size to your weight, but it can become bigger or smaller if the forces are unbalanced, causing you to feel heavier or lighter.

The diagram below shows the forces acting on a person moving in a vertical circle when they are at the top and bottom of the circle.

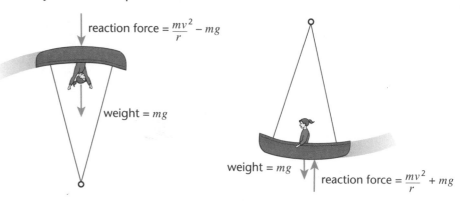

The forces on a person moving in a vertical circle

At the bottom of the circle:
• The person feels heavier than usual as the size of the normal reaction force is equal to the person's weight plus the centripetal force required to maintain circular motion.

At the top of the circle:
• The person may feel lighter than usual as the person's weight is providing part of the centripetal force required to maintain circular motion. The rest comes from the normal reaction force, which is less than the person's weight if $mv^2/r < 2mg$.

• If the speed of the ride is such that the required centripetal force is the same as the person's weight then he or she feels 'weightless' as the normal reaction force is zero.

Progress check

1 a Convert into radians:
  i 90°
  ii 720°

b An object is displaced through an angle of 8π radians.
  Explain why it has returned to its original position.

2 A stone is tied to a string and whirled round in a horizontal circle.
  a What is providing the centripetal force?
  b The stone moves at a horizontal speed of 3.5 m s⁻¹.
    The radius of the circle is 2.0 m.
    Calculate the centripetal acceleration.
  c Calculate the size of the centripetal force if the mass of the stone is 0.25 kg.

c 1.53 N
b 6.125 m s⁻²
2 a The tension in the string.
b It has travelled four complete revolutions.
ii 4π
1 a i π/2

Sample question and model answer

In this question, take the value of g, free fall acceleration, to be 10 m s^{-2}.

A toy car, mass 60 g, is at rest at the foot of a slope.

A second car, mass 90 g, which is moving at a speed of 1.20 m s^{-1} hits the back of the stationary car and the cars stick together.

1.20 m s^{-1}

When objects stick together, there is always some energy transfer from kinetic energy to internal energy, as some deformation occurs.

However, total energy is always conserved, no matter what type of collision.

(a) (i) What type of collision occurs?   [1]

Inelastic   1 mark

(ii) Which of the following quantities are conserved in this collision?   [1]

kinetic energy   momentum   total energy

Momentum and total energy.   1 mark

Always write down the formula that you intend to use, and each step in the working. Otherwise, a simple error such as using the wrong mass results in no marks.

(b) (i) Calculate the momentum of the larger car before the collision.   [3]

momentum = mass × velocity   1 mark
         = 0.090 kg × 1.20 m s^{-1} 1 mark
         = 0.108 kg m s^{-1} 1 mark

(ii) Calculate the combined speed of the cars after the collision.   [3]

The first mark here is for correct transposition of the equation.

velocity = momentum ÷ mass   1 mark
         = 0.108 kg m s^{-1} ÷ 0.150 kg   1 mark
         = 0.72 m s^{-1} 1 mark

(c) Calculate the vertical height of the cars' centre of mass when they come to rest on the slope.   [5]

It is not essential to include the units at each stage of the calculation, but it is good practice and you MUST include the correct unit with your answer to each part.

As the cars rise up the slope, kinetic energy is transferred to gravitational potential energy   1 mark
$\frac{1}{2}mv^2 = mg\Delta h$ 1 mark
$\Delta h = v^2 \div 2g$ 1 mark
        = $(0.72 \text{ m s}^{-1})^2 \div (2 \times 10 \text{ m s}^{-2})$ 1 mark
        = 0.026 m   1 mark

Practice examination questions

Throughout this section, take the value of *g*, free fall acceleration, to be 10 m s⁻².

1

A child of mass 40 kg sits on a swing. An adult pulls the swing back, raising the child's centre of mass through a vertical height of 0.60 m. This is shown in the diagram.

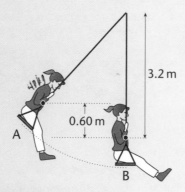

(a) Calculate the increase in gravitational potential energy when the child moves from position B to position A. $mgh = 40 \times 0.6 \times 10$ [3]
$$= 240 \text{ J}$$

(b) The child is released from position A.
Calculate her speed as she passes through position B. [4]
$240 = \frac{1}{2} \times 40 \times v^2$ $= \frac{240}{20}$ $= \sqrt{6}$ m/s

(c) Explain why there must be a resultant (unbalanced) force on the child as she swings through position B and state the direction of this force. [4]

(d) Calculate the size of the resultant force that acts on the child as she swings through position B. [3]

(e) One force that acts on the child at position B is the upward push of the swing. Write a description of the other force that acts on the child. [2]

(f) Write down the value of the upward push of the swing when the child is:
(i) stationary at B.
(ii) moving through B. [2]

2

The diagram shows two 'vehicles' on an air track approaching each other. After impact they stick together.

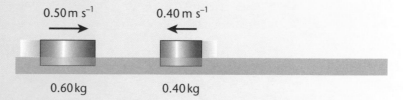

(a) Calculate the total momentum of the vehicles before the collision. [3]

(b) Calculate the velocity of the vehicles after the collision. State the direction of motion. [3]

(c) By performing appropriate calculations, explain whether the collision is elastic or inelastic. [3]

3

A communications satellite in a geostationary orbit (orbit time = 24 hours) goes round the Earth at a distance of 4.24×10^7 m from the centre of the Earth.

(a) Calculate the orbital speed of the communications satellite. [3]

(b) Calculate the centripetal acceleration of the satellite and state its direction. [4]

(c) What force causes this acceleration? [2]

4

A tennis ball has a mass of 60 g. A ball travelling at a speed of 15 m s^{-1} hits a player's racket and rebounds with a speed of 24 m s^{-1}. The ball is in contact with the racket for a time of 12 ms.

(a) Calculate the change in momentum of the ball. [3]

(b) Calculate the average force exerted by the racket on the ball. [3]

(c) Calculate the average force exerted by the ball on the racket. [1]

(d) Describe the energy transfers that take place while the ball is in contact with the racket. [3]

5

The diagram shows a tractor being used to pull a log.

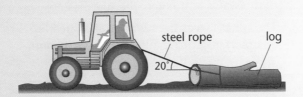

The tension in the steel rope is 600 N.

(a) Calculate the horizontal component of the forwards force on the log. [2]

(b) What other horizontal force acts on the log and what is the size of this force? [2]

(c) Calculate the work done on the log when it is pulled a distance of 250 m. [3]

(d) What happens to the energy used to pull the log? [2]

6

A car of mass 800 kg travels around a circular corner of radius 100 m at a speed of 14 m s⁻¹.

(a) Calculate the centripetal acceleration of the car. [3]

(b) Calculate the size of the force needed to cause this acceleration. [3]

(c) (i) Write a description of the force that causes this acceleration. [2]

(ii) What is the other force that makes up the pair of forces in the sense of Newton's third law? [2]

(d) Suggest why, in wet or icy conditions, cars sometimes leave the road when travelling round a bend. [2]

7

A jet aircraft has a mass of 500 000 kg when fully laden.
It accelerates from rest to a take-off speed of 60 m s⁻¹ in 20 s.

(a) Calculate the momentum of the aircraft as it leaves the ground. [2]

(b) Calculate the mean force required to impart this momentum to the aircraft. [2]

(c) Suggest why the force produced by the engines during take-off is greater than the answer to (b). [2]

(d) How does the principle of conservation of momentum apply to this event? [2]

8

In a hydroelectric power station, water falls at the rate of 3.2×10^5 kg s⁻¹ through a vertical height of 180 m before entering a turbine.

(a) Calculate the loss in gravitational potential energy of the water each second. [3]

(b) Assuming that all this energy is transferred to kinetic energy, calculate the speed of the water as it enters the turbines. [3]

(c) The water leaves the turbines at a speed of 3 m s⁻¹. Calculate the maximum energy transfer to the turbine each second. [2]

Chapter 2
Waves

The following topics are covered in this chapter:

- *Oscillations, resonance and damping*
- *Simple harmonic motion*
- *Wave properties*
- *Complex wave behaviour*
- *Quantum phenomena*

2.1 Oscillations, resonance and damping

After studying this section you should be able to:

- *explain the difference between a free and a forced vibration*
- *describe resonance and how it occurs*
- *distinguish between a damped oscillation and an undamped one and give examples of each*

LEARNING SUMMARY

Oscillations

AQA A	M4	OCR A	M4
AQA B	M4	OCR B	M4
EDEXCEL A	M4	NICCEA	M4
EDEXCEL B	M4	WJEC	M4

The terms 'oscillation' and 'vibration' have the same meaning. They are both used to describe a regular to-and-fro movement.

An **oscillation** is a repetitive to-and-fro movement. Loudspeaker cones oscillate, as do swings in a children's playground. However, there is an important difference between these oscillations.

The loudspeaker cone is an example of a **forced vibration**; the cone is forced to vibrate at the frequency of the current that passes in the coil. The amplitude of vibration depends on the size of the current. When reproducing speech or music, the frequency of vibration is constantly changing.

A child's swing, like a string on a piano or a guitar is an example of a **free vibration**. Once displaced from its normal position, it vibrates at its own **natural frequency**. The natural frequency of vibration of an object is the frequency it vibrates at when it is displaced from its normal, or rest, position and released. For a swing, this frequency depends only on the length of the chain or rope; in the case of a stringed instrument it is also affected by the tension and mass per unit length of the string.

The vibration of particles in a solid is a free vibration. Like a swing, the energy of the particles changes from kinetic to potential (stored) energy.

- In a free vibration there is a constant interchange between potential (stored) energy and kinetic energy.

- A swing has zero kinetic energy and maximum gravitational potential energy when at its greatest displacement; the kinetic energy is at a maximum and the gravitational potential energy at a minimum when the displacement is zero. This is shown in the diagram.

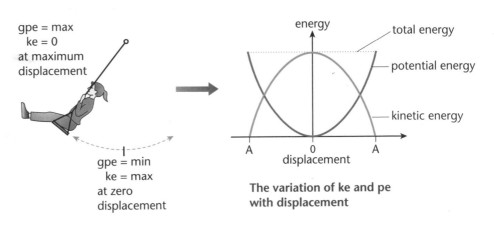

gpe = max
ke = 0
at maximum
displacement

gpe = min
ke = max
at zero
displacement

The variation of ke and pe with displacement

Resonance

AQA A	M4	OCR A	M4
AQA B	M4	OCR B	M4
EDEXCEL A	M4	NICCEA	M4
EDEXCEL B	M4	WJEC	M4

There are many everyday examples of resonance. You may have noticed windows vibrating due to the sound from a bus waiting at a stop. Electrical resonance is used to tune in a radio or television to a particular station.

After a swing is given a push, it vibrates at its natural frequency. To make it swing higher, subsequent pushes need to coincide with the vibrations of the swing. This is an example of **resonance**, the large amplitude oscillation that occurs when the frequency of a forced vibration is equal to the natural frequency of the vibrating object.

The diagram shows how a vibration generator can be used to force a string to vibrate at any frequency, and the large amplitude vibration that occurs when the driving frequency coincides with the natural frequency of the string, also called its **resonant frequency**. At this frequency the string is able to absorb sufficient energy from the vibration generator to replace the energy lost due to resistive forces.

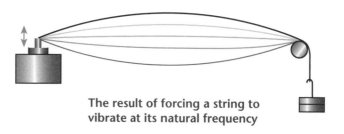

The result of forcing a string to vibrate at its natural frequency

If the frequency of the forced oscillation is increased slowly from zero, the amplitude of vibration of the string starts to increase at a frequency below the resonant frequency. It has its maximum value at this frequency and it then decreases again as the frequency continues to increase. The variation of amplitude around the resonant frequency is shown below.

The sharpness of the peak depends on the resistive forces acting. The greater the resistive force, the smaller the amplitude of vibration.

Vibration generator frequency

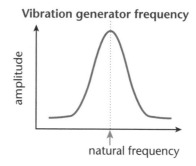

natural frequency

- Microwave cookers heat food by causing water molecules in the food to resonate, absorbing the energy from the microwaves.

- Magnetic resonance imaging (MRI) is used in medicine to produce a detailed picture of the inside of the body by making the atomic nuclei resonate.

These are some everyday examples of resonance.

- Resonance in mechanical systems can be a nuisance or a danger. When the frequency of the engine vibrations in a vehicle matches the natural frequency of the exhaust the result is a rattle as the exhaust vibrates against the body of the vehicle.

- Resonance of parts of aircraft can cause failure due to excessive stress and several helicopter crashes have been attributed to resonance of the pilot's eyeballs resulting in the pilot being unable to see overhead power lines.

Damping

Just like linear motion, all mechanical oscillations are subject to resistive forces. The effect of resistive forces in removing energy from a vibrating object is known as damping.

A string or mass on a spring vibrating in air is lightly damped as the main resistive force is air resistance. The effect of light damping is to gradually reduce the energy, and therefore amplitude, of the vibrating object.

The body of a motor vehicle is connected to the wheels by springs. When the vehicle goes over a bump in the road the springs compress, giving the passengers a smoother ride. If motor vehicles were lightly damped they would continue to oscillate after going over a bump. Shock absorbers increase the resistive force, so that when the body of the vehicle has been displaced it returns to its original position without oscillating. A motor vehicle is critically damped so that it returns to the normal position in the minimum time to avoid vibrations that could cause 'car-sickness' and possibly loss of control by the driver.

> A car mechanic tests the shock absorbers by pushing down on the car bonnet. If the car oscillates, the shock absorbers need replacing.

Very large resistive forces result in heavy damping. Imagine a mass on a spring suspended so that the mass is in a viscous liquid such as syrup. If the mass is displaced, the force opposing its movement is very large and it takes a long time to return to the normal position.

> The effect of damping is to transfer the energy from the vibrating object to heat in the surroundings.

One effect of damping is to decrease the effects of resonance. The greater the damping, the smaller the amplitude of vibration at the resonant frequency. This is very important in mechanical structures such as bridges which can resonate due to the effects of wind. Resonance caused the suspension bridge built over the Tacoma Narrows, in America, to collapse in 1939. Since then, engineers have designed suspension bridges with shock absorbers built into the suspension to absorb the energy and prevent excessive vibration.

Progress check

1 Describe the difference between a free vibration and a forced vibration.
2 Explain why a washing machine sometimes vibrates violently when the drum is spinning.
3 Explain how a free vibration is affected by the amount of damping.

3 Damping removes energy from a vibrating object. Increasing the damping reduces the amplitude of vibration and may stop any vibration from occurring.
2 The washing machine has a natural frequency of vibration. If the drum rotates at this frequency it causes the machine to resonate.
1 A free vibration occurs when an object is displaced and it vibrates at its natural frequency. A forced vibration can be at any frequency and is caused by an external oscillating force.

2.2 Simple harmonic motion

After studying this section you should be able to:

- explain what is meant by simple harmonic motion
- apply the equations of simple harmonic motion
- sketch graphs showing the change in displacement, velocity and acceleration in simple harmonic motion

LEARNING SUMMARY

What is simple harmonic motion?

AQA A	M4	OCR A	M4
AQA B	M4	OCR B	M4
EDEXCEL A	M4	NICCEA	M4
EDEXCEL B	M4	WJEC	M4

> The resultant force on an object in simple harmonic motion is a restoring force since it is always directed towards the equilibrium position that the object has been displaced from.

As the name implies, it is the simplest kind of oscillatory motion. One set of equations can be used to describe and predict the movement of any object whose motion is simple harmonic.

The motion of a vibrating object is simple harmonic if:

- its acceleration is proportional to its displacement
- its acceleration and displacement are in opposite directions.

The second bullet point means that the acceleration, and therefore the resultant force, always acts towards the equilibrium position, where the displacement is zero.

Common examples of simple harmonic motion (often abbreviated to shm) include the oscillations of a simple pendulum (provided the amplitude is small, $\leq 10°$) and those of a mass suspended vertically on a spring.

The diagram shows the size of the acceleration of a simple pendulum and a mass on a spring when they are given a small displacement, x, from the equilibrium position.

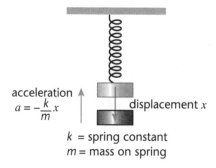

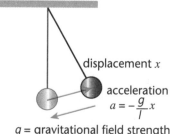

acceleration
$a = -\dfrac{k}{m}x$ displacement x

k = spring constant
m = mass on spring

displacement x

acceleration
$a = -\dfrac{g}{l}x$

g = gravitational field strength
l = length of pendulum

> k, the spring constant, is defined as the force per unit extension
> $k = f/x$

In each case:

- the numerical value of the acceleration is equal to a constant multiplied by the displacement, showing that acceleration is proportional to displacement
- the negative value of the acceleration shows that it is in the opposite direction to the displacement, since acceleration and displacement are both vector quantities.

Frequency of simple harmonic motion

AQA A	M4	OCR A	M4
AQA B	M4	OCR B	M4
EDEXCEL A	M4	NICCEA	M4
EDEXCEL B	M4	WJEC	M4

The constants in the equations for the acceleration of the mass on a spring (constant = k/m) and the pendulum bob (constant = g/l) give useful information about the frequency and time period of the oscillation.

> Since $(2\pi f)^2$ is always positive, a and x have the opposite signs so they are always in opposite directions.

> **KEY POINT**
>
> The acceleration of any object whose motion is simple harmonic is related to the frequency of vibration by the equation:
> $$a = -(2\pi f)^2 x$$
> where a is the acceleration, f is the frequency and x is the displacement.

Key points from AS

- **The difference between distance and displacement**
  Revise AS section 1.5

A graph of acceleration against displacement is therefore a straight line through the origin, as shown in the diagram. The gradient of the graph has a negative value and is numerically equal to $(2\pi f)^2$.

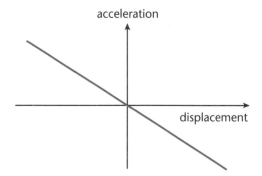

The frequency, f, and time period, T, are related to the constants in the equations. For a mass on a spring:

$$(2\pi f)^2 = k/m$$

So:

> $f = \frac{1}{2\pi} \sqrt{\frac{k}{m}}$ and $T = 2\pi \sqrt{\frac{m}{k}}$ **KEY POINT**

For a simple pendulum:

$$(2\pi f)^2 = g/l$$

So:

> $f = \frac{1}{2\pi} \sqrt{\frac{g}{l}}$ and $T = 2\pi \sqrt{\frac{l}{g}}$ **KEY POINT**

These equations for frequency and time period show that:

- increasing the mass on a spring causes the frequency to decrease and the time period to increase
- for a given mass, the stiffer the spring the higher the frequency of oscillation
- the frequency and time period of a pendulum do not depend on the mass of the pendulum bob
- increasing the length of a pendulum causes the frequency to decrease and the time period to increase.

The relationship between the frequency, f, and the time period, T, is $f = 1/T$ so each of these quantities is the reciprocal of the other.

The spring constant, k, is a measure of the stiffness of a spring.

Displacement and time

AQA A	M4	OCR A	M4
AQA B	M4	OCR B	M4
EDEXCEL A	M4	NICCEA	M4
EDEXCEL B	M4	WJEC	M4

The displacement of an object in simple harmonic motion varies sinusoidally with time. This means that a displacement–time graph has the shape of a sine or cosine curve, and the displacement at any time, t, can be written in terms of a sine or cosine function.

> The displacement of an object in simple harmonic motion can be calculated using either of the equations:
> $$x = A \sin 2\pi ft$$
> $$x = A \cos 2\pi ft$$
> where A is the amplitude of the motion. **KEY POINT**

The symbol x_0 is also used as an alternative to A to represent amplitude.

The graphs below show the variation of displacement with time for a simple harmonic motion.

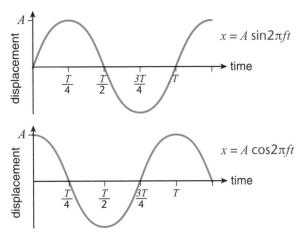

Sine and cosine are identical functions with a 90°, or π/2 radians, phase difference.

Which equation and graph apply to a simple harmonic motion depends on the displacement when the time is zero.

- If the displacement is zero at time $t = 0$, then the sine applies.
- If the displacement is at the maximum at time $t = 0$, then the cosine applies.

Displacement, velocity and acceleration

AQA A	M4	OCR A	M4
AQA B	M4	OCR B	M4
EDEXCEL A	M4	NICCEA	M4
EDEXCEL B	M4	WJEC	M4

These bullet points apply to any motion.

Displacement, velocity and acceleration are linked graphically because:

- the gradient of a displacement–time graph represents velocity
- the gradient of a velocity–time graph represents acceleration.

The following diagram shows the variation of all three variables for a simple harmonic motion where the variation of the displacement with time is represented by a sine function.

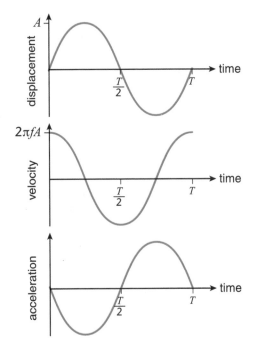

These statements about the acceleration follow from the definition of simple harmonic motion.

These graphs show that:

- when the displacement is zero the velocity is at a maximum and the acceleration is zero
- when the displacement has its maximum value the velocity is zero and the acceleration is a maximum in the opposite direction.

Knowledge of the frequency and amplitude of a simple harmonic motion enables the maximum speed and the velocity at any displacement to be calculated.

The ± symbol is used because at any value of the displacement the velocity could be in either direction.

The maximum speed of an object in simple harmonic motion is:
$$v = 2\pi fA$$
The relationship between velocity and displacement is:
$$v = 2\pi \sqrt{(A^2 - x^2)}$$

Progress check

When a 100 g mass is attached to the bottom of a vertical spring it causes it to extend by 5.0 cm.

a Calculate the value of the spring constant, k.
  Assume $g = 10$ N kg^{-1}.
  The mass is pulled down a further 3 cm and released.
b Calculate the frequency and time period of the resulting oscillation.
c **i** What is the value of the amplitude of the oscillation?
  ii Calculate the maximum speed of the mass.

ii 0.42 m s^{-1}
c i 3 cm
b 2.25 Hz and 0.44 s
a 20 N m^{-1}

2.3 Wave properties

After studying this section you should be able to:

- describe the difference between a longitudinal and a transverse wave
- recall and apply the wave equation to calculate speed, frequency and wavelength
- explain the meaning of polarisation and intensity

Wave motion

AQA A	M4	OCR A	AS
AQA B	AS	NICCEA	AS
EDEXCEL A	M4	WJEC	AS
EDEXCEL B	AS		

The waves on a water surface are almost transverse. The particles move in an elliptical path.

Waves are used to transfer a signal or energy. They do this without an accompanying flow of material, although some waves, such as sound, can only be transmitted by the particles of a substance.

All waves consist of **vibrations** or **oscillations**. Sound and other compression waves are classified as **longitudinal** because the vibrations of the particles carrying the wave are along, or parallel to, the direction of wave travel. All electromagnetic waves are **transverse**; the vibrations of the electric and magnetic fields in these waves are at right angles to the direction of wave travel.

Waves transmitted through the body of water, or any other liquid, are longitudinal.

The diagram shows a longitudinal wave being transmitted by a spring and a transverse wave on a rope.

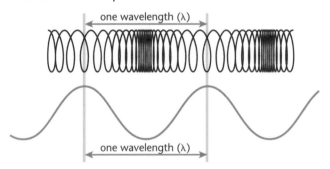

one wavelength (λ)

one wavelength (λ)

Wave measurements

AQA A	M4	OCR A	AS
AQA B	AS	OCR B	AS
EDEXCEL A	M4	NICCEA	AS
EDEXCEL B	AS	WJEC	AS

Most waves that you come across every day are progressive. Waves that are not progressive are discussed in the section 'complex wave behaviour'.

These measurements apply to all **progressive waves**. A progressive wave is one that has a profile that moves through space.

- **wavelength** (symbol λ) is the length of one complete cycle; a compression (squash) and rarefaction (stretch) in the case of a longitudinal wave, a peak and a trough in the case of a transverse wave (see diagram above).
- **amplitude** (symbol a) is the maximum displacement from the mean position (see diagram page 51).
- **frequency** (symbol f) is the number of vibrations per second, measured in hertz (Hz)
- **speed** (symbol v) is the speed at which the profile moves through space
- **period** (symbol T) is the time taken for one vibration to occur. It is related to frequency by the equation $T = 1/f$

For all waves, the wavelength, speed and frequency are related by the equation:

$$\text{speed} = \text{frequency} \times \text{wavelength}$$
$$v = f \times \lambda$$

Wave graphs

AQA A	M4	OCR A	AS
EDEXCEL A	M4	NICCEA	AS
EDEXCEL B	AS		

The definition of the size of an angle in radians is *arc length ÷ radius,* so it follows that an angle of $360° = 2\pi r/r = 2\pi$ radians.

A displacement–position or displacement–time graph does not distinguish between a transverse wave and a longitudinal one.

Try substituting values of *n*, for *n* = 0, 1, 2 etc. into the expression for 'an odd number of half wavelengths'. This will help you to understand the meaning of the phrase.

A common error is to misinterpret a CRO display as a graph of displacement against distance. This leads to a horizontal measurement, which represents time, being interpreted as a distance.

Both transverse and longitudinal waves can be represented by graphs of displacement against position. The position along a wavefront can be measured in two ways:

- as a distance from a point in space
- as an angle.

In angular measure, one complete cycle of the wave is represented by an angle of 360°, or 2π radians. This is shown in the diagram.

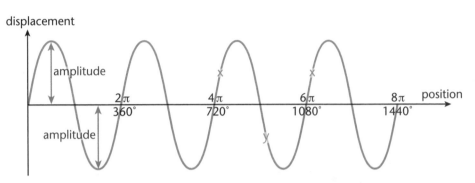

Using angular measure is a convenient way of comparing the **phase** of different parts of a wave. Two points on a wave are **in phase** if they have the same displacement and velocity. For progressive waves, this is only true for points exactly *n* wavelengths or $2\pi n$ radians or $360n°$ degrees apart, where *n* is an integer. Points marked x on the diagram above are in phase.

The point marked y is exactly **out of phase** with the points x as it has the opposite displacement and its velocity has the same value but in the opposite direction. This applies to any two points separated by an odd number of half wavelengths (i.e. 1/2, 3/2, 5/2 wavelengths, etc.). This can be expressed algebraically as $(2n + 1)\lambda/2$ or $(2n + 1)\pi$, where *n* is an integer.

The **phase difference** between two points on a wave describes their relative displacement and velocity and is normally expressed in degrees or wavelengths. A phase difference of $\lambda/2$ or 180° or π radians describes two points separated by an odd number of half wavelengths (i.e. exactly out of phase), while a phase difference of $\lambda/4$ or 90° or $\pi/2$ radians compares two points that are separated by one-quarter, five-quarters, nine-quarters of a wavelength, etc.

A **cathode ray oscilloscope** (CRO) can be used to display a different type of graph that represents wave motion. Rather than showing the displacement along a wave, the CRO plots the displacement of one point on the wave against time. The display on the CRO looks the same as that in the diagram, but the horizontal axis represents time rather than position on the wave.

To measure the **period** and **frequency** of a wave from a CRO display, the time-base control, which controls the rate at which the dot sweeps the screen horizontally, needs to be in the 'cal' (calibration) position. The diagram shows a CRO display of a sound wave, with the time–base set to 2 ms cm⁻¹.

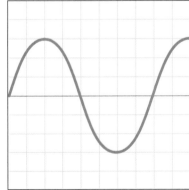

One cycle of the wave occupies a distance of 8 cm on the screen, so the period, $T = 8 \times 2$ ms = 16 ms = 1.6×10^{-2} s.

The frequency of the wave, $f = 1/T = 1 \div 1.6 \times 10^{-2}$ s = 62.5 Hz.

Polarisation

AQA A	M4	OCR A	AS
AQA B	AS	NICCEA	AS
EDEXCEL A	M4	WJEC	AS
EDEXCEL B	AS		

Polarising material is used in some types of sunglasses and in filters for camera lenses. It cuts out reflected light from a flat surface such as an expanse of water.
Can you work out how it does this?

In order to receive a television or radio broadcast, the receiving aerial has to be lined up with that of the transmitter. This is because radio waves broadcast from aerials are polarised, the vibrations are only in one plane. For radio and television broadcasts the plane of polarisation is usually either vertical or horizontal.

Light waves are not normally polarised, and nor are radio waves from stars. In an unpolarised wave the vibrations are in all planes at right angles to the direction of travel. Polarisation does not apply to longitudinal waves.

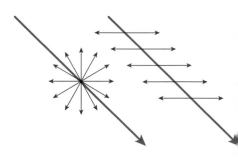

The diagram opposite shows the vibrations in polarised and unpolarised waves. Light waves can become polarised as they pass through some materials and they are partially polarised when reflected.

Intensity

AQA B	AS	EDEXCEL A	M4

The signal strength of a radio broadcast decreases the further away the receiver is from the transmitting aerial. In a similar way, a sound appears fainter and a light seems dimmer further away from their origins. This is due to energy becoming spread over a wider area as the wave travels away from its source.

The intensity with which a wave is received depends on the power incident on the area of the detector.

A point source such as a star radiates energy in all directions. In this case the variation of intensity with distance follows an inverse square pattern, with:

$$I \propto 1/r^2$$

> **KEY POINT**
>
> Intensity = power per unit area measured at right angles to the direction of travel.
>
> $$I = P/A$$
>
> Intensity is measured in W m^{-2}.

The diagram shows the sound spreading out from a loudspeaker. The ear that is further away detects a quieter sound because the power is spread over a wider area, so less enters the ear.

Progress check

1 A television station broadcasts on a wavelength of 0.60 m. The speed of the waves is 3.0×10^8 m s^{-1}. Calculate the frequency of the waves.

2 Two waves have a phase difference of 60°. What is the phase difference in radians?

3 Explain why sound cannot be polarised.

3 Sound is a longitudinal wave, so there are no vibrations at right angles to the direction of travel.
2 $\pi/3$.
1 5.0×10^8 Hz.

2.4 Complex wave behaviour

After studying this section you should be able to:

- describe how the diffraction of a wave at a gap depends on the wavelength and size of the gap
- explain how wave superposition causes interference patterns and describe the conditions for the patterns to be observable
- explain the formation of a stationary wave

LEARNING SUMMARY

Diffraction

AQA A	M4	OCR B	AS
AQA B	AS	NICCEA	AS
EDEXCEL A	M4	WJEC	AS
OCR A	AS		

Our everyday experience is that sound can travel round corners, but light travels in straight lines. The spreading out of waves as they pass through openings or the edges of obstacles is called diffraction. All waves can be diffracted, but the effect is more noticeable with long wavelength waves such as sound and radio waves than with short wavelength waves such as light and X-rays.

The diagrams show what happens when water waves pass through gaps of different sizes.

The amount of spreading when the wave has passed through the gap depends on the relative sizes of the gap and the wavelength.

When answering questions about diffraction, always emphasise the size of the gap compared to the wavelength.

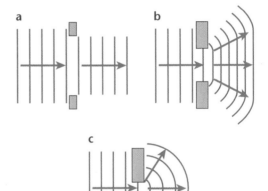

- For a gap that is many wavelengths wide, see diagram **a**, no detectable spreading takes place.

- Some spreading occurs when waves pass through a gap that is several wavelengths wide; diagram **b**.

- The maximum amount of spreading corresponds to a gap that is the same size as the wavelength as in diagram **c**.

Diagram **a** models what happens when light (wavelength approximately 5×10^{-7} m) passes through a doorway, and diagram **c** models sound (wavelength approximately 1 m) passing through the same doorway.

Diffraction explains why you can 'hear' round corners but you cannot 'see' round corners.

Diffraction also explains why long wavelength radio broadcasts can be detected in the shadows of hills and buildings, but short wavelength broadcasts cannot. Diffraction is also important in the design of loudspeakers to maximise the spreading for sounds of all wavelengths and in satellite transmissions to minimise the diffraction that occurs.

Superposition and interference of sound

AQA A	M4	OCR A	AS
AQA B	AS	OCR B	AS
EDEXCEL A	M4	NICCEA	AS
EDEXCEL B	AS	WJEC	AS

The superposition of waves occurs when two or more waves cross. This is happening all the time, but the effects are so short-lived that we seldom notice them. One of the easiest effects of superposition to observe is two-source interference. This can be demonstrated with sound, light, surface water waves and 3 cm electromagnetic waves. The diagram shows the variation in the loudness of the sound in front of two loudspeakers vibrating in phase.

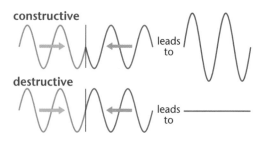

Superposition of sound waves

An observer walking along a line parallel to the loudspeakers notices that:

- there are regions where the sound is much louder than from one loudspeaker alone
- there are regions where the sound is barely audible
- in between these, the intensity of the sound varies from loud to quiet.

This is explained using the principle of superposition, which describes how two waves can reinforce each other or cancel each other out.

The term 'vector sum' here means that the directions are taken into account, so that a positive displacement can be cancelled by a negative one.

This means that displacements in the same direction cause reinforcement, while displacements in opposite directions cause cancellation. The diagram below shows how a loud sound and no sound can be produced when two waves of equal amplitude meet at a point.

- If the waves are exactly in step (in phase) then constructive interference takes place, resulting in a loud sound
- If the waves are exactly out of step (out of phase) then destructive interference takes place, resulting in no sound.

Conditions for interference to be observed

AQA A	M4	OCR B	AS
AQA B	AS	NICCEA	AS
EDEXCEL A	M4	WJEC	AS
OCR A	AS		

The amplitudes do not need to be the same, but they should be comparable for the interference to be observed. If one wave has a much greater amplitude than the other then the effects of cancellation and reinforcement will not be noticeable.

The waves do not need to have the same amplitude for interference to take place. The effect of one wave having a greater amplitude than the other is that the destructive interference is not total. This happens when the observer is closer to one loudspeaker than to the other.

The diagram on page 55 represents the waves from two sources vibrating in phase. The solid and broken lines could represent peaks and troughs in the case of water waves or compressions and rarefactions in the case of sound waves. Some lines of constructive interference (c) and destructive interference (d) have been marked in.

If the sources in the diagram are in phase, then at any point equidistant from both sources the waves arrive in phase, giving constructive interference. This is the case for points along the central c line in this diagram. At all other points in the interference pattern, there is a path difference, i.e. the wave from one source has travelled further than that from the other source.

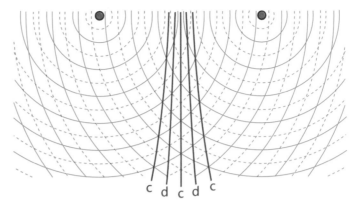

Constructive and destructive interference from two wave sources

Try drawing similar patterns and investigating the effect of changing the wavelength and the separation of the sources.

> **KEY POINT**
>
> A difference in the path length causes a difference in the phase of two waves arriving at a point. The relationship between phase difference and path difference is:
>
> $$\Delta(\text{phase}) = \frac{2\pi\Delta(\text{path})}{\lambda}$$

The conditions can be written as $n\lambda$ path difference for constructive interference and $(2n+1)\lambda/2$ path difference for destructive interference, where n is an integer.

- **Constructive interference** takes place when the path difference is a whole number of wavelengths.
- **Destructive interference** takes place when the path difference is one and a half wavelengths, two and a half wavelengths, etc.

This is shown in the diagram.

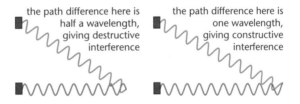

Clearly, for an interference pattern to be observed, the sources must not only be coherent but also of the same type.

These conditions for constructive and destructive interference only apply to two sources that are in phase. If the phase of one of the oscillators in the diagram is reversed so that the sources are in antiphase, then the conditions for constructive and destructive interference are also reversed. The effect is to shift the interference pattern so that lines of constructive interference (c) become destructive and *vice versa*. Any value of phase difference between the two sources gives an interference pattern that is stationary, provided that the phase difference is not changing. Two sources with a **fixed phase difference** are said to be **coherent**. Coherent sources must have the same wavelength and frequency.

Interference of light

AQA A	M4	OCR B	AS
AQA B	AS	NICCEA	AS
EDEXCEL A	M4	WJEC	AS
OCR A	AS		

Interference patterns are easy to set up with water waves and sound because two sources can be driven from the same oscillator, giving coherence. Because of the way in which light is emitted in random bursts of energy from a source, it is not possible to have two separate sources that are coherent. Instead, diffraction is used to obtain two identical copies of the light from a single source. This is done by illuminating two narrow slits with a lamp filament that is parallel to the slits so that the same wavefronts arrive at each slit. A suitable arrangement is shown in the diagram overleaf.

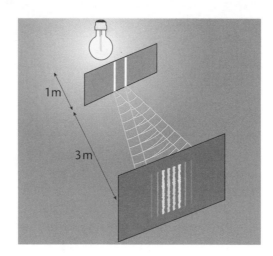

A colour filter can be used between the lamp and the slits to reduce the range of wavelengths interfering, but this also reduces the intensity of the interference pattern.

Bright and dark fringes on the screen are due to constructive and destructive interference between the two overlapping beams of light. The separation of the fringes depends on:

- the separation of the slits
- the wavelength of the light
- the distance between the slits and the screen.

This formula enables the wavelength of light to be measured from a simple experiment. The fringe spacing, x, should be obtained by measuring the separation of as many fringes as are visible and dividing by the number of fringes.

> **KEY POINT**
>
> The separation of the fringes is related to the other variables by the formula:
>
> $$\text{fringe spacing} = \frac{\text{wavelength} \times \text{distance from slits to screen}}{\text{slit separation}}$$
> $$x = \lambda D / a$$
>
> where x is the distance between adjacent bright (or dark) fringes
> λ is the wavelength of the light
> D is the distance between the slits and the screen
> a is the distance between the slits.

As the wavelength of light is very small, the slits need to be close together and separated from the screen by a large distance for the interference fringes to be seen.

The diffraction grating

A diffraction grating consists of a large number of narrow slits that are ruled very close together on a glass slide. Diffraction takes place at each of the slits and results in an interference pattern that enables precise measurements of wavelength. The photograph shows the interference pattern obtained from a cadmium light source.

This photograph shows an emission spectrum, the wavelengths of light emitted by the source.

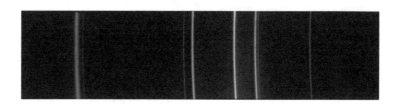

The bright lines in the photograph are due to constructive interference taking place between light diffracted at thousands of slits. The diagram on page 57 shows the path difference when the diffracted light is viewed at an angle θ to the normal.

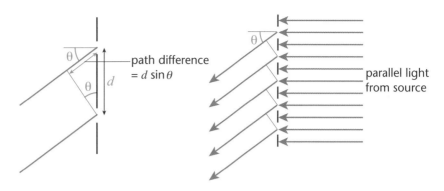

Bright lines are seen when waves from all the slits interfere constructively. For this to happen the path difference shown in the diagram has to be equal to a whole number of wavelengths.

When *n* = 0, all wavelengths interfere constructively, so the bright central band consists of all wavelengths that are incident on the grating.

> **KEY POINT**
>
> The condition for constructive interference to occur when light is diffracted at a diffraction grating is:
>
> $$d \sin \theta = n\lambda$$
>
> where *d* is the distance between adjacent slits
> θ is the angle between the direction of the light and the normal
> *n* is an integer
> λ is the wavelength of the light.

When *n* = 1 the bright bands are known as the **first order spectrum**. The number of spectra that are visible depends on the separation of the slits.

Diffraction gratings are used to analyse the wavelengths of light emitted from gas discharge tubes and the light received from stars.

Stationary (standing) waves

AQA A	M4	OCR A	AS
AQA B	AS	OCR B	AS
EDEXCEL A	M4	NICCEA	AS
EDEXCEL B	AS	WJEC	AS

A progressive wave is one whose profile moves through space, whereas a stationary wave has a static profile. The waves on the vibrating strings and in the vibrating air columns of musical instruments are stationary waves.

Stationary waves are caused by the superposition of two waves of the same wavelength travelling in opposite directions. They often arise when a wave is reflected at a boundary, but the waves can come from two separate coherent sources. The diagram shows the vibrations of a stationary wave on a string.

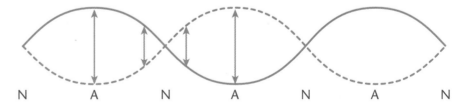

Stationary waves differ from progressive waves in a number of respects:

All points on a progressive wave vibrate with the same amplitude, but each point is only in phase with points precisely *n* wavelengths away.

- there is no flow of energy along a stationary wave, although stationary waves often radiate energy
- within each loop of a stationary wave, all particles vibrate in phase and exactly out of phase (π radians or 180° phase difference) with the particles in adjacent loops
- the amplitude of vibration varies with position in the loop
- there are **nodes** (points marked N in the diagram), where the displacement is always zero and **antinodes** (marked A in the diagram) which vibrate with the maximum amplitude.

A convenient way of measuring the wavelength of a progressive wave is to use it to set up a stationary wave, for example by superposing the wave on its reflection from a barrier, and measuring the distance between a number of successive nodes or antinodes.

The wavelength of a stationary wave is twice the distance between two adjacent nodes or antinodes. The diagram below shows how a stationary wave (shown with a solid line) is formed from two progressive waves (shown with broken lines) travelling in opposite directions. In (a) the waves interfere constructively, so each point on the wave has its maximum displacement. In (b) each progressive wave has moved one quarter of a wavelength; the interference is now destructive, resulting in no displacement at all points on the wave.

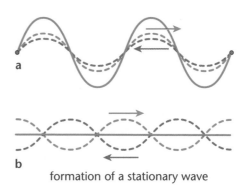

formation of a stationary wave

Progress check

1 Two waves arriving at a point interfere constructively. What are the possible values of the phase difference between the waves?

2 Two loudspeakers are placed 2.5 m apart and face in the same direction. They vibrate in phase, emitting sound with a wavelength of 0.3 m. An observer walks along a line parallel to the line joining the loudspeakers, 5.0 m in front of them.
   a Calculate the distance between two adjacent loud sounds being heard.
   b What difference would there be if the phase of one of the loudspeakers was reversed?

3 How is diffraction used to produce a two-source interference pattern using light?

3 Diffraction is used to produce two identical copies of a single light source by passing a light through two parallel narrow slits.
   b The loud sounds would become quiet and the quiet sounds would become loud.
2 a 0.60 m
1 Zero or a whole number multiple of 2π radians.

2.5 Quantum phenomena

After studying this section you should be able to:

- describe the main features of the electromagnetic spectrum
- explain how the photoelectric effect gives evidence for the particulate nature of electromagnetic radiation
- explain line spectra and relate the spectral lines to energy level transitions

LEARNING SUMMARY

The spectrum

AQA A	AS	NICCEA	AS
EDEXCEL A	M4	WJEC	AS
OCR A	AS		

The speed of all electromagnetic waves in a vacuum is 2.999×10^8 m s^{-1}. The value 3.00×10^8 m s^{-1} is normally used for both a vacuum and air.

The different changes of speed when light consisting of a range of wavelengths is refracted is responsible for dispersion, the splitting of light into a spectrum.

Microwaves are given here as a separate part of the spectrum, although they can be considered as short-wavelength radio waves.

The electromagnetic spectrum consists of a whole family of waves, with some similarities and some differences in their behaviour.

Similar properties include:

- they consist of electric and magnetic fields oscillating at right angles to each other
- they travel at the same speed in a vacuum
- they are transverse waves, so they can be polarised
- they show the same pattern of behaviour in reflection, refraction, interference and diffraction.

The differences include:

- the shorter wavelength waves undergo a greater change of speed when being refracted
- wave-like properties such as diffraction and interference are more readily observable in the behaviour of the longer wavelength waves.

The diagram shows the range of wavelengths and frequencies of the waves that make up the spectrum.

| frequency / Hz | 10^{20} | | 10^{17} | | 10^{14} | | 10^{11} | | 10^{8} | | 10^{5} |

gamma rays   ultraviolet   infra-red   radio waves

X-rays   light   microwaves

| wavelength / m | 10^{-12} | | 10^{-9} | | 10^{-6} | | 10^{-3} | | 1 | | 10^{3} |

In general, the name given to an electromagnetic wave depends on its wavelength or frequency. Notable exceptions to this are X-rays and gamma rays, whose ranges of wavelength and frequency overlap. The difference here is in the origin of the waves, an X-ray is emitted when a high-speed electron is suddenly brought to rest and a gamma ray is emitted from an excited nucleus, usually following alpha or beta emission.

The table shows typical wavelengths, the origins of the waves and the uses of the main parts of the electromagnetic spectrum.

name of radiation	typical wavelength/m	how produced	used for
gamma	1×10^{-12}	an excited nucleus releasing energy in radioactive decay	tracing the flow of fluids and treating cancer
X-rays	1×10^{-10}	high-speed electrons being stopped by a target	seeing inside the body
ultraviolet	1×10^{-8}	very hot objects and passing electricity through gases	suntan and security marking

light	5×10^{-7}	hot objects and passing electricity through gases	vision and photography
infra-red	1×10^{-5}	warm and hot objects	heating and cooking
microwave	1×10^{-1}	microwave diode or oscillating electrons in an aerial	communications and cooking
radio	1×10^{2}	oscillating electrons in an aerial	communications

All objects give out infra-red radiation, no matter what the temperature. The hotter the object, the more power is emitted and the greater the range of wavelengths.

Waves or particles

AQA A	AS	OCR A	AS
AQA B	M4	OCR B	AS
EDEXCEL A	M4	NICCEA A	AS
EDEXCEL B	AS	NICCEA B	AS

The **threshold wavelength**, λ_0, is the wavelength of the waves that have the threshold frequency.
$$\lambda_0 = c \div f_0$$

The photoelectric effect can be demonstrated using a zinc plate connected to a gold leaf electroscope. An ultraviolet lamp discharges a negatively-charged plate but has no effect on a positively-charged plate.

The word quantum refers to the smallest amount of a quantity that can exist. A quantum of electromagnetic radiation is the smallest amount of energy of that frequency.

The **photoelectric effect** provides evidence that electromagnetic waves have a particle-like behaviour which is more pronounced at the short-wavelength end of the spectrum. In the photoelectric effect electrons are emitted from a metal surface when it absorbs electromagnetic radiation.

The results of photoelectricity experiments show that:

- there is no emission of electrons below a certain frequency, called the **threshold frequency**, f_0, which is different for different metals
- above this frequency, electrons are emitted with a range of kinetic energies up to a maximum, $(\frac{1}{2}mv^2)_{max}$
- increasing the frequency of the radiation causes an increase in the maximum kinetic energy of the emitted electrons, but has no effect on the photoelectric current, i.e. the rate of emission of electrons
- increasing the intensity of the radiation has no effect if the frequency is below the threshold frequency; for frequencies above the threshold it causes an increase in the photoelectric current, so the electrons are emitted at a greater rate.

The wave model cannot explain this behaviour; if electromagnetic radiation is a continuous stream of energy then radiation of all frequencies should cause photoelectric emission. It should only be a matter of time for an electron to absorb enough energy to be able to escape from the attractive forces of the positive ions in the metal.

The explanation for the photoelectric effect relies on the concept of a **photon**, a quantum or packet of energy. We picture electromagnetic radiation as short bursts of energy, the energy of a photon depending on its frequency.

A lamp emits random bursts of energy. Each burst is a photon, a quantum of radiation.

The relationship between the energy, E of a photon, or quantum of electromagnetic radiation, and its frequency, f, is:
$$E = hf$$
where h is Planck's constant and has the value 6.63×10^{-34} J s.

KEY POINT

The energy of a photon can be measured in either joules or **electronvolts**. The electronvolt is a much smaller unit than the joule.

One electron volt (1 eV) is the energy transfer when an electron moves through a potential difference of 1 volt.
$$1 \text{ eV} = 1.60 \times 10^{-19} \text{ J}$$

KEY POINT

The conversion factor for changing energies in eV to energies in joules is 1.60×10^{-19} J eV^{-1}

The work function is the **minimum** energy needed to liberate an electron from a metal. Some electrons need more than this amount of energy.

Radiation below the threshold frequency, f_0, no matter how intense, does not cause any emission of electrons.

Einstein's explanation of photoelectric emission is:

- An electron needs to absorb a minimum amount of energy to escape from a metal. This minimum amount of energy is a property of the metal and is called the work function, ϕ.
- If the photons of the incident radiation have energy hf less than ϕ then there is no emission of electrons.
- Emission becomes just possible when $hf = \phi$.
- For photons with energy greater than ϕ, the electrons emitted have a range of kinetic energies, those with the maximum energy being the ones that needed the minimum energy to escape.
- Increasing the intensity of the radiation increases the number of photons incident each second. This causes a greater rate of emission of electrons, but does not affect their maximum kinetic energy.

> **KEY POINT**
>
> Einstein's photoelectric equation relates the maximum kinetic energy of the emitted electrons to the work function and the energy of each photon:
> $$hf = \phi + (\tfrac{1}{2}mv^2)_{max}$$

At the threshold frequency, the minimum frequency that can cause emission from a given metal, $(\tfrac{1}{2}mv^2)_{max}$ is zero and so the equation becomes $hf_0 = \phi$.

Stopping the emission

A stopping potential is positive compared to earth.

The work function of a metal is the minimum photon energy that causes photoelectric emission. This minimum energy can be increased by applying a positive potential, V, to the metal. An electron now needs an additional amount of energy eV (in joules, or numerically equal to V in eV) to escape.

> **KEY POINT**
>
> The **stopping potential** V_s, is the potential applied to a metal that just stops photoelectric emission. At the stopping potential:
> $$hf = \phi + eV_s$$

Particles or waves

AQA A	AS	OCR A	AS
AQA B	M4	OCR B	AS
EDEXCEL A	M4	NICCEA	AS
EDEXCEL B	M4	WJEC	AS

Try calculating the de Broglie (pronounced de Broy) wavelength of a moving snooker ball. Is it possible for the ball to show wave-like behaviour?

If waves can show particle-like behaviour in photoelectric emission, can particles also behave as waves? Snooker balls bounce off cushions in the same way that light bounces off a mirror, so reflection is not a test for wave-like or particle-like behaviour. Diffraction and interference are properties unique to waves, so particles can be said to have a wave-like behaviour if they show these properties. All particles have an associated wavelength called the de Broglie wavelength:

> **KEY POINT**
>
> The wavelength, λ, of a particle is related to its momentum, p, by the de Broglie equation:
> $$\lambda = h/p = h/mv$$
> where h is the Planck constant.

The de Broglie wavelength of such an electron is of the order of 1×10^{-10} m.

An electron that has been accelerated through a potential difference of a few hundred volts has a wavelength similar to that of X-rays and gamma rays.

This wavelength is also similar to the spacing of the atoms in crystalline materials, so these materials provide suitable sized 'gaps' to cause diffraction.

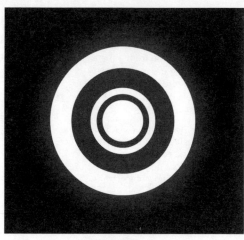

A diffraction pattern formed by passing a beam of electrons through graphite

Diffraction patterns formed by a beam of electrons after passing through thin foil or graphite show a set of 'bright' and 'dark' rings on photographic film, similar to those formed by X-ray diffraction.

Electrons can also be made to interfere when two coherent beams overlap. They produce an interference pattern similar to that of light, but on a much smaller scale.

Other 'particles' such as protons and neutrons also show wave-like behaviour.

> You may have seen this pattern formed on a fluorescent screen in a vacuum tube.

> Particles and waves are the models that we use to describe and explain physical phenomena. It is not surprising that the real world does not fit neatly into our models.

There are two separate models of how matter behaves. The particle model explains such phenomena as ionisation and photoelectricity, while the wave model explains interference and diffraction. It is not appropriate to classify matter as 'waves' or 'particles' as photons and electrons can fit either model, depending on the circumstances.

Line spectra

AQA A	M6A	OCR A	M5.1
AQA B	M4	OCR B	M5
EDEXCEL A	M4	NICCEA	AS
EDEXCEL B	AS	WJEC	AS

> Gamma rays are the exception because they are a result of the nucleus losing energy.

With the exception of gamma rays, the emission of electromagnetic radiation is associated with electrons losing energy. A hot solid can radiate the whole range of wavelengths through the infra-red and part of the visible spectrum, the extent of the range depending on its temperature.

When an electric current is passed through an ionised gas, only a small number of wavelengths are emitted. The wavelengths are characteristic of the gas used and are called a line spectrum.

The diagram below shows some of the lines present in the hydrogen spectrum and their wavelengths.

> Not all of these lines lie in the visible spectrum, which extends from 0.4 µm to 0.65 µm.

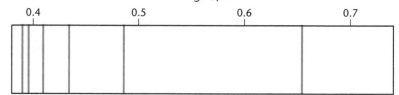

wavelength/µm

The existence of line spectra provides evidence that the electrons in orbit around a nucleus can only have certain values of energy, the values being characteristic of an atom. Energy can only be emitted or absorbed in amounts that correspond to the differences between these allowed values.

An energy level diagram shows the amounts of energy that an electron can have. The diagram opposite is an energy level diagram for hydrogen.

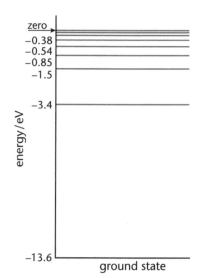

Energy levels in a hydrogen atom

Note that on an energy level diagram:

- the energies are measured relative to a zero that represents the energy of an electron at rest outside the atom, i.e. one that is just 'free'.
- an orbiting electron has less energy than a free electron, so it has negative energy relative to the zero

Movement of an electron to the ground state results in the emission of a photon with an energy in excess of 10 eV. Photons with this amount of energy give spectral lines in the ultraviolet region of the spectrum.

- an electron with the minimum possible energy is in the ground state; higher energy levels are called excited states.

Energy is emitted in the form of a photon when an electron moves from an excited state to a lower energy level. The energy of the photon is equal to the difference in the values of the energy levels.

> **KEY POINT**
>
> When an electron moves from an energy level E_1 to a lower energy level E_2, the energy of the photon emitted is given by
>
> $$hf = E_1 - E_2$$

The lines of the emission spectrum often appear black in the absorption spectrum. The photons of these energies are absorbed as the electrons move to more excited states, and then released when the electrons lose energy. The emitted photons are radiated in all directions, so very little energy is detected in any one direction.

For example, an electron moving from an energy level with -0.38 eV of energy to one with -0.85 eV loses energy equal to $(-0.38 - -0.85)$ eV $= 0.47$ eV. This corresponds to emitting a photon of frequency 1.13×10^{14} Hz, which lies in the infra-red part of the electromagnetic spectrum.

Electrons can gain energy by absorbing photons. As with emission, the only photons that can be absorbed are those that correspond to allowed movements, or transitions, of electrons. An absorption spectrum is produced by shining white light through a sample of a gaseous element. The spectrum that emerges is the full spectrum with the element's emission spectrum 'missing' or of low intensity. This is due to the electrons absorbing photons of just the right energy to allow them to move to a more excited state.

Electrons in atoms

The simple model of a hydrogen atom pictures a proton with a single electron in orbit. Since the electron is continually accelerating it ought to be radiating energy and spiralling in towards the nucleus. The fact that it does not do this is because of its wave-like behaviour. Each allowed orbit of the electron corresponds to a standing wave, with all the energy being contained within the wave and none being radiated.

Progress check

1 The work function of potassium is 2.84 eV. Calculate:
   a the minimum frequency of radiation that causes photoelectric emission.
   b the maximum energy of the emitted electrons when ultraviolet radiation of frequency 8.40×10^{14} Hz is absorbed by potassium.

2 Calculate the de Broglie wavelength of a neutron, mass $= 1.67 \times 10^{-27}$ kg, travelling at a speed of 200 m s^{-1}.
   $h = 6.63 \times 10^{-34}$ J s

3 Calculate the energy of a photon that causes the excitation of an electron in a hydrogen atom from the ground state (-13.6 eV) to an excited state with an energy of -1.5 eV:
   a in eV
   b in J.

3 b 1.94 × 10⁻¹⁸ J
3 a 12.1 eV
2 1.99 × 10⁻⁹ m
1 b 1.03 × 10⁻¹⁹ J
1 a 6.85 × 10¹⁴ Hz

Sample question and model answer

The diagram shows two similar loudspeakers that are driven from a common signal generator and vibrate in phase. The loudspeakers are 0.50 m apart and emit a note of frequency 1500 Hz.

An observer walks along a line 3.0 m from the loudspeakers, as shown in the diagram. He hears a loud sound at A, which decreases to a minimum at B and then increases in intensity as he walks towards C.

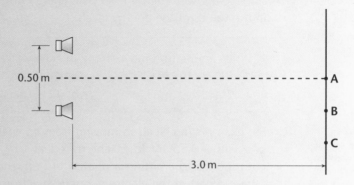

(a) Explain why the observer detects a loud sound at A. [2]

A is equidistant from both loudspeakers, so as the waves are emitted in phase, they must arrive at A in phase.

The vibrations arrive at A in phase (1 mark) so they interfere constructively (1 mark).

(b) Explain why the loudness of the sound detected decreases as the observer walks from A to B and is a minimum at B. [3]

The phase difference changes gradually from being totally in phase at A to being totally out of phase at B.

The phase difference increases from A to B (1 mark); the sound becomes quieter due to less constructive/more destructive interference (1 mark) and is a minimum at B where the vibrations are out of phase (1 mark).

(c) The speed of sound in air is 340 m s⁻¹.

Calculate the wavelength of the sound emitted by the loudspeakers. [3]

$\lambda = v/f$ 1 mark
$= 340$ m s⁻¹ ÷ 1500 Hz   1 mark
$= 0.227$ m   1 mark

This relationship is valid for all examples of two–source interference, provided that the separation of the sources is small compared to the distance between the sources and the observer.

(d) Calculate the distance between A and C. [3]

separation of maxima, $x = \lambda D/a$ 1 mark
$= 0.227$ m × 3.0 m ÷ 0.50 m   1 mark
$= 1.36$ m   1 mark

(e) The connections to one of the loudspeakers are reversed. What difference does the observer notice as he walks from A to C? Explain why he notices these differences. [2]

The points of maximum intensity become minima and vice versa (1 mark) as the loudspeakers are now vibrating out of phase (1 mark).

Practice examination questions

Throughout this section use the following values of the electronic charge and Planck's constant:

$e = -1.60 \times 10^{-19}$ C

$h = 6.63 \times 10^{-34}$ J s

1

(a) Explain the meaning of the expression *natural frequency of vibration.* [2]

(b) In what circumstances does an object resonate? [2]

(c) The speed of sound in air is 340 m s^{-1}.

Sound waves with a wavelength of 0.15 m cause a glass tumbler to vibrate at its natural frequency.

Calculate the natural frequency of vibration of the tumbler. [3]

2

A mass is suspended on a spring. After it is given a small displacement its equation of motion is:

$$a = -kx/m$$

(a) State the meaning of the symbols used in this equation. [4]

(b) The diagram shows how the displacement of the mass changes during one cycle of oscillation.

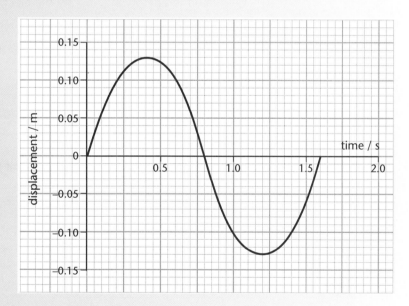

Calculate:

    (i) the amplitude of the oscillation [1]

    (ii) the frequency of the oscillation. [2]

(c) The mass on the spring is 0.50 kg. Calculate the spring constant. [3]

(d) (i) Calculate the maximum kinetic energy of the mass. [3]

    (ii) At what point on its motion does the mass have its maximum kinetic energy? [1]

3

The diagram represents some of the energy levels in a hydrogen atom.

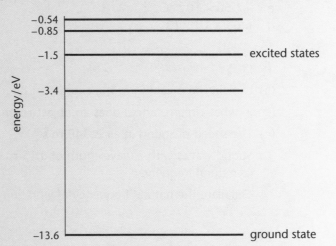

(a) (i) What is meant by the **ground state**? [1]

(ii) Explain why an electron in the ground state has a negative amount of energy. [2]

(b) An electron undergoes a transition from an energy level of –0.85 eV to the ground state.

(i) Calculate the energy, in J, lost by the electron. [2]

(ii) Calculate the frequency of the photon that is emitted. [2]

(iii) In which part of the electromagnetic spectrum is the radiation emitted as a result of this transition? [1]

(iv) Give one danger and one use of radiation in this part of the spectrum. [2]

4

In a demonstration of two-source interference, a source of electromagnetic waves of wavelength 2.8 centimetres is directed at a metal barrier. There are two gaps in the barrier. The arrangement is shown in the diagram.

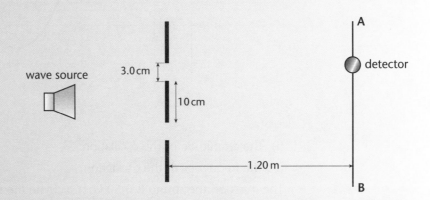

(a) (i) Suggest why a single wave source is directed at two gaps in the barrier. [2]

(ii) Explain why 3.0 cm is a suitable size for the gaps. [2]

(b) Where is the strongest signal detected?
Explain why this is the strongest signal. [3]

(c) The detector is moved along the line AB.
What is the distance between the points where the strongest and weakest signals are detected? [4]

(d) The wave source is replaced by a lamp that emits light of one wavelength only.
The detector is replaced by a white screen placed along the line AB.
Explain why no interference pattern is observed. [2]

5

Einstein's equation for photoelectric emission is:

$$hf = \Phi + (\tfrac{1}{2}mv^2)_{max}$$

(a) (i) State the meaning of each term. [3]

(ii) What is the significance of the subscript 'max'? [1]

(b) Electromagnetic radiation of frequency 2.02×10^{15} Hz is directed at a metal surface. The maximum kinetic energy of the electrons emitted is 4.05×10^{-19} J. The intensity of the radiation is 4.5 μW m^{-2}.

Calculate:

(i) the energy of each photon [2]

(ii) the number of photons reaching the surface each second if the area is 2.5×10^{-6} m^2 [ ]

(iii) The work function of the metal. Give your answer in eV. [3]

6

The graph shows how the displacement of a pendulum bob changes over one cycle of its motion.

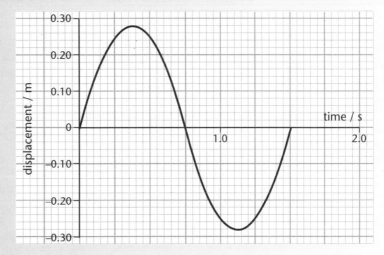

The equation that represents the variation of displacement with time is $x = A \sin 2\pi ft$.

(a) Write down the values of A and f. [2]

(b) Calculate the maximum velocity of the pendulum bob. [1]

(c) Calculate the velocity of the pendulum bob when its displacement is 0.20 m. [2]

Chapter 3
Fields

The following topics are covered in this chapter:

- Gravitational fields
- Electric fields
- Capacitors

- Magnetic fields
- Circular orbits
- Electromagnetic induction

3.1 Gravitational fields

After studying this section you should be able to:

- understand the concept of a field
- recall and use the relationship that describes the gravitational force between two masses
- describe the Earth's gravitational field and explain how the field strength varies with distance from the centre of the Earth

LEARNING SUMMARY

Fields

AQA A	M4	OCR A	M4
AQA B	M5	OCR B	M4
EDEXCEL A	M5	NICCEA	M4
EDEXCEL B	M5	WJEC	M5

Radiant energy such as light has mass and so is affected by gravitational fields.

A field is a region of space where forces are exerted on objects with certain properties. In this and subsequent sections three types of field are considered:

- **gravitational fields** affect anything that has mass
- **electric fields** affect anything that has charge
- **magnetic fields** affect permanent magnets and electric currents.

These three types of field have many similar properties and some important differences. There are key definitions and concepts that are common to all three types of field.

Gravitational fields

AQA A	M4	OCR A	M4
AQA B	M5	OCR B	M4
EDEXCEL A	M5	NICCEA	M4
EDEXCEL B	M5	WJEC	M5

The mass of the Earth is about 6×10^{24} kg.

Newton realised that all objects with mass attract each other. This seems surprising, since any two objects placed close together on a desktop do not immediately move together. The attractive force between them is tiny, and very much smaller than the frictional forces that oppose their motion.

Gravitational attractive forces between two objects only affect their motion when at least one of the objects is very massive. This explains why we are aware of the force that attracts us and other objects towards the Earth – the Earth is very massive.

The diagram represents the Earth's gravitational field. The lines show the direction of the force that acts on a mass that is within the field.

This diagram shows that:

- gravitational forces are always attractive – the Earth cannot repel any objects
- the Earth's gravitational pull acts towards the centre of the Earth
- the Earth's gravitational field is radial; the field lines become less concentrated with increasing distance from the Earth.

The force exerted on an object in a gravitational field depends on its position. The less concentrated the field lines, the smaller the force. If the gravitational field strength at any point is known, then the size of the force can be calculated.

> Gravitational field strength is a vector quantity: its direction is towards the object that causes the field.

KEY POINT

The gravitational field strength (g) at any point in a gravitational field is *the force per unit mass* at that point:
$$g = F/m$$
Close to the Earth's surface, g has the value of 9.81 N kg^{-1}, though the value of 10 N kg^{-1} is often used in calculations.

Universal gravitation

AQA A	M4	OCR A	M4
AQA B	M5	OCR B	M4
EDEXCEL A	M5	NICCEA	M4
EDEXCEL B	M5	WJEC	M5

In studying gravitation, Newton concluded that the gravitational attractive force that exists between any two masses:

- is proportional to each of the masses
- is inversely proportional to the square of their distances apart.

> A point mass is one that has a radial field, like that of the Earth.

KEY POINT

Newton's law of gravitation describes the gravitational force between two point masses. It can be written as:
$$F = \frac{Gm_1m_2}{r^2}$$
where G is the universal gravitational constant and has the value
6.7 × 10^{-11} N m^2 kg^{-2}
where m_1 and m_2 are the values of the masses and r is the separation of the centres of mass.

> Remember that two objects attract *each other* with equal-sized forces acting in opposite directions.

Although the Earth is a large object, on the scale of the Universe it can be considered to be a point mass. The gravitational field strength at its centre is zero, since attractive forces pull equally in all directions. Beyond the surface of the Earth, the gravitational force on an object decreases with increasing distance. When the distance is measured from the centre of the Earth, the size of the force follows an inverse square law; doubling the distance from the centre of the Earth decreases the force to one quarter of the original value. The variation of force with distance from the centre of the Earth is shown in the diagram.

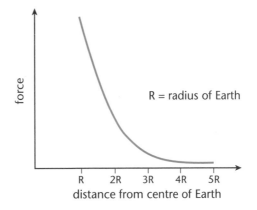

69

g and G

> A negative sign is sometimes used on the right-hand side of this equation, following the convention that attractive forces are given negative values and repulsive forces positive values.

Newton's law of gravitation can be used to work out the value of the force between any two objects. It can also be used to calculate the strength of the gravitational field due to a spherical mass such as the Earth or the Sun.

> **KEY POINT**
>
> A small object, mass m, placed within the gravitational field of the Earth, mass M, experiences a force, F, given by
>
> $$F = \frac{GMm}{r^2}$$
>
> where r is the separation of the centres of mass of the object and the Earth.

It follows from the definition of gravitational field strength as the *force per unit mass* that the field strength at that point, g, is related to the mass of the Earth by the expression:

> **KEY POINT**
>
> $$g = \frac{F}{m} = \frac{GM}{r^2}$$

Gravitational field strength is a property of any point in a field. It can be given a value whether or not a mass is placed at that point. Like gravitational force, beyond the surface of the Earth the value of g follows an inverse square law. A graph of g against distance from the centre of the Earth has the same shape as that shown in the diagram on page 69.

> The radius of the Earth is about 6.4×10^6 m, so you would have to go much higher than aircraft-flying height for g to change by 1%.

Because the inverse square law applies to values of g when the distance is measured from the centre of the Earth, there is little change in its value close to the Earth's surface. Even when flying in an aircraft at a height of 10 000 m, the change in distance from the centre of the Earth is minimal, so there is no noticeable change in g.

The same symbol, g, is used to represent:

- gravitational field strength
- free-fall acceleration.

These are not two separate quantities, but two different names for the same quantity. Gravitational field strength, g, is defined as the force per unit mass, $g = F/m$. From Newton's second law and the definition of the newton, free-fall acceleration, g, is also equal to the gravitational force per unit mass. The units of gravitational field strength, N kg^{-1}, and free-fall acceleration, m s^{-2}, are also equivalent.

Potential and potential energy

> On an absolute scale of measurement, zero must be the smallest possible value and '20 units' must be twice as much as '10 units'.

When an object changes its position relative to the Earth, there is a change in potential energy given by $\Delta E_p = mg\Delta h$. It is not possible to place an absolute value on the potential energy of any object when h is measured relative to the surface of the Earth. Two similar objects placed at the top and bottom of a hill have different values of potential energy, but relative to the ground the potential energy is zero for both objects, see diagram on the next page.

Absolute values of potential energy are measured relative to infinity. In this context, infinity means 'at a distance from the Earth where its gravitational field strength is so small as to be negligible'.

The car at the top of the hill has more potential energy than the one at the bottom, but relative to ground level they both have zero.

Using this reference point:

* all objects at infinity have the same amount of potential energy, zero
* any object closer than infinity has a negative amount of potential energy, since it would need to acquire energy in order to reach infinity and have zero energy.

Just as gravitational field strength is used to place a value on the gravitational force that would be experienced by a mass at any point in a gravitational field, the concept of gravitational potential is used to give a value for the potential energy.

> **KEY POINT**
>
> The gravitational potential at a point in a gravitational field is the potential energy per unit mass placed at that point, measured relative to infinity.

So if the potential at any point in a field is known, the potential energy of a mass placed at that point can be calculated by multiplying the potential by the mass.

Calculating potential and potential energy

AQA A	M4	OCRA B	M4
AQA B	M4	WJEC	M5
EDEXCEL A	M5		

When an object is within the gravitational field of a planet, it has a negative amount of potential energy measured relative to infinity. The amount of potential energy depends on:

* the mass of the object
* the mass of the planet

* the distance between the centres of mass of the object and the planet.

> **KEY POINT**
>
> The gravitational potential energy measured relative to infinity of a mass, m, placed within the gravitational field of a spherical mass M can be calculated using:
>
> $$E_p = -\frac{GMm}{r}$$
>
> where r is the distance between the centres of mass and G is the universal gravitational constant.
>
> Gravitational potential energy is measured in joules (J).

Since gravitational potential is the gravitational potential energy per unit mass placed at a point in a field, it follows that:

> **KEY POINT**
>
> Gravitational potential, V, is given by the relationship:
>
> $$V = \frac{E_p}{m} = -\frac{GM}{r}$$
>
> Gravitational potential is measured in J kg^{-1}.

The relationship between potential and potential energy is similar to that between gravitational force and gravitational field strength:

- potential and field strength are properties of a point in a field
- potential energy and force are the corresponding properties of a mass placed within a field.

Equipotential surfaces

AQA B M5   OCR B M4
EDEXCEL A M5

For a satellite in an elliptical orbit, there is an interchange between kinetic and potential energy as it travels around the Earth.

The potential energy of a satellite in a circular orbit around the Earth remains constant provided that its distance from the centre of the Earth does not change. To move to a higher or lower orbit the satellite must gain or lose energy. The satellite travels along an equipotential surface, the spherical shape consisting of all points at the same potential.

The diagram shows the spacing of equipotential surfaces around the Earth. The surfaces are drawn at equal differences of potential.

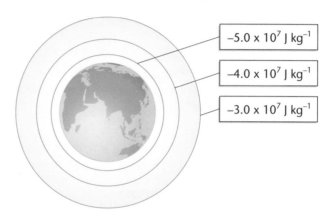

The diagram shows that:

- equipotential surfaces around a spherical mass are also spherical
- the spacing between the equipotential surfaces increases with increasing distance from the centre of the Earth.

For a satellite to move from an orbit where the potential is -4.0×10^7 J kg⁻¹ to one where the potential is -3.0×10^7 J kg⁻¹, it needs to gain 1.0×10^7 J of gravitational potential energy for each kilogram of satellite. It does this by firing the rocket engines, transferring energy from its fuel supply.

To move to a lower orbit, a rocket can lose engines by firing the rocket engines 'backwards' so that the exhaust gases are expelled in the direction of motion.

Calculating potential

AQA A M4   OCR B M4
AQA B M5   WJEC M5
EDEXCEL A M5

The value of the gravitational potential at a point in a gravitational field depends on:

- the mass of the object causing the field
- the distance from the centre of mass of this object.

> **KEY POINT**
>
> The gravitational potential, V, due to a spherical mass, M, at a distance r from its centre of mass is given by:
> $$V = -\frac{GM}{r}$$

The diagram above shows that the rate of change of potential with distance, the potential gradient, decreases with increasing distance from the Earth.

As the potential gradient decreases, so does the gravitational field strength.

The concept of potential gradient is similar to that of the gradient of a hill or slope. The steeper the slope, the greater the acceleration of an object free to move down it.

The relationship between gravitational field strength and potential gradient is:

$$g = - \text{(potential gradient)} \quad \text{or} \quad g = -\frac{\Delta V}{\Delta x}$$

where ΔV is the change in potential over a small distance Δx.

Progress check

1 The radius of the Earth is 6.4×10^6 m and the gravitational field strength at its surface is 10 N kg^{-1}.
   At what height above the surface of the Earth is the gravitational field strength equal to 2.5 N kg^{-1}?

2 Two 2.5 kg masses are placed with their centres 10 cm apart.
   Calculate the size of the gravitational attractive force between them.
   $G = 6.7 \times 10^{-11}$ N m^2 kg^{-2}.

3 The mass of the Moon is 7.4×10^{22} kg and its radius is 1.7×10^6 m.
   Using the value of G from Q2, calculate the value of free-fall acceleration at the Moon's surface.

3 1.7 m s^{-2}
2 4.2×10^{-8} N
1 6.4×10^6 m

3.2 Electric fields

After studying this section you should be able to:

- *describe the electric field due to a point charge and between two charged parallel plates*
- *calculate the force on a charge in an electric field*
- *compare gravitational and electric fields*

Charging up

EDEXCEL A   M5

Electrostatic phenomena such as attraction and repulsion are due to the forces between two charged objects. Similar charges repel and opposite charges attract.

Transfer of charge happens whenever two objects in physical contact move relative to each other. It is caused by electrons leaving one surface and joining another. This results in objects having one of two types of charge:

- an object that gains electrons has a negative charge
- an object that loses electrons has a positive charge.

In many cases any imbalance of charge on an object is removed by movement of electrons to and from the ground, but if at least one of the objects is a good insulator charge can build up.

A balloon is easily charged by rubbing but it is not possible to charge a hand-held metal rod since it is immediately discharged by electrons passing through the person holding it.

The **quantum**, or smallest unit, of charge is that carried by an electron. All quantities of charge must be a whole number multiple of $e = -1.6 \times 10^{-19}$ C.

The electric field

AQA A	M4	OCR A	M4
AQA B	M5	OCR B	M5
EDEXCEL A	M5	NICCEA	M4
EDEXCEL B	M4	WJEC	M5

Unlike gravitational fields, which can only exert attractive forces, electric fields can attract or repel objects that are charged. When drawing field lines that represent the forces due to a charged object, the arrows show the direction of the force on a positive charge.

The diagram shows the electric fields due to a point charge and between a pair of oppositely charged parallel plates.

A pair of parallel plates become oppositely charged when connected to the positive and negative terminals of a d.c. supply.

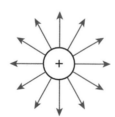
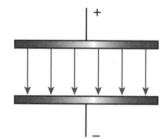

The field due to the point charge is radial, it decreases in strength with increasing distance from the charge. That between the parallel plates is uniform, it maintains a constant strength at all points between the plates.

Coulomb investigated the size of the force between two point charges and concluded that the force is:

- proportional to each of the charges
- inversely proportional to the square of their distances apart.

Permittivity is a measure of the extent to which the medium reinforces the electric field. Water has a high permittivity due to its molecules being polarised.

KEY POINT

Coulomb's law states that the force between two point charges is given by

$$F = \frac{kQ_1 Q_2}{r^2}$$

where Q_1 and Q_2 represent the values of the charges, r is their distance apart and k has the value, in air or a vacuum, of $\frac{1}{4\pi\varepsilon_0} = 9.0 \times 10^9$ N m^2 C^{-2}.

The constant ε_0 is called the permittivity of free space.

Field strength

AQA A	M4	OCR A	M4
AQA B	M5	OCR B	M5
EDEXCEL A	M5	NICCEA	M4
EDEXCEL B	M4	WJEC	M5

Electric field strength:

- is defined *as the force per unit positive charge acting on a small charge placed within the field*

The test charge has to be small enough to have no effect on the field.

- is measured in N C^{-1}.

Coulomb's law can be used to express the field strength due to a point charge Q. Since the force between a charge Q and a small charge q placed with the field of Q is given by $F = \frac{kQq}{r^2}$, it follows that:

KEY POINT

The electric field strength, E, due to a point charge Q is given by the expression:

$$E = \frac{kQ}{r^2}$$

In a radial field, the field strength follows an inverse square law. This can be seen by the way in which the field lines spread out from a point charge. In a uniform field, like the one between two oppositely charged parallel plates, the field lines maintain a constant separation. The value of the electric field strength in a uniform field does not change.

This gives an alternative unit for electric field strength, V m^{-1}, which is equivalent to the N C^{-1}.

KEY POINT

The electric field strength between two oppositely charged parallel plates is given by the expression:

$$E = \frac{V}{d}$$

where V is the potential difference between the plates and d is the separation of the plates.

Potential

AQA A	M4	OCR B	M5
AQA B	M5	WJEC	M5
EDEXCEL A	M5		

The potential difference between the parallel plates shown in the diagram on page 74 represents the energy transfer per coulomb when charge moves between them. A charge q moving between the plates would gain or lose energy Vq.

In a uniform field the potential changes by equal amounts for equal changes in distance.

The diagram here shows the variation in potential between two oppositely charged plates.

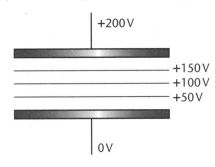

Potential could be measured relative to the upper plate, in which case the potentials would have negative values.

In this example:

- potential has been measured relative to the lower plate, which has been given the value 0
- the equipotential lines are parallel to the plates
- an equipotential surface, joining points all at the same potential, is simply a surface drawn parallel to the plates.

Potential in a radial field

AQA A	M4	OCR B	M5
AQA B	M5	WJEC	M5
EDEXCEL A	M5		

Unlike the potential in the gravitational field of a point mass, the electric potential in the field of a point charge is positive, since work has to be done to move a positive charge from infinity to any point in the field.

Like potential in a gravitational field, absolute potential in an electric field is measured relative to infinity. Infinity does not need to be a great distance, since it refers to a point where the field strength is negligibly small.

> **KEY POINT**
>
> The electric potential at a point, V, is the work done per unit positive charge in bringing a small charge from infinity to that point.
>
> In a radial field, $$V = \frac{1}{4\pi\varepsilon_0}\frac{Q}{r}$$

Electron beams

AQA A	M4	EDEXCEL A	M5
AQA B	M5	NICCEA	M5

Computer monitors, televisions and cathode ray oscilloscopes all use beams of electrons to produce a picture. The beam is produced by an electron gun, a device that accelerates and focuses the electrons given off by a hot wire. The diagram below shows an electron gun.

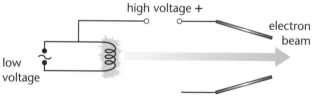

An electron gun

The filament is the cathode, or negative terminal. The anode is usually cylindrical and is connected to the positive terminal of the high voltage supply.

The low voltage supply heats the filament, causing it to emit electrons. The electrons are accelerated by the high voltage supply, gaining kinetic energy as they move towards the positively charged anode. Since the potential difference between the cathode and the anode represents the energy transfer per coulomb of charge, the kinetic energy of the accelerated electrons can be calculated.

> **KEY POINT**
>
> When an electron is accelerated through a potential difference V it gains kinetic energy
> $$E_k = eV$$
> $$\tfrac{1}{2}m_e v^2 = eV$$
> where m_e is the mass of an electron and e is the electronic charge.

Deflecting a beam of electrons

AQA A	M4	EDEXCEL B	M4
AQA B	M5	NICCEA	M5
EDEXCEL A	M5		

In a cathode ray oscilloscope, vertical deflection of the electron beam is achieved by passing the beam between a pair of oppositely charged parallel plates. The effect of this can be seen by studying the path of a beam of electrons in a deflection tube.

When passing between the plates, the electrons have a constant speed in the direction parallel to the plates. Perpendicular to the plates, the force on each electron is equal to *Ee*, so they accelerate in this direction. The consequent motion of the electrons is similar to that of a particle projected horizontally on the Earth; in each case the result is a parabolic path, shown in the diagram.

Electric fields are also used to deflect other charged particles, for example the ink droplets in an inkjet printer.

The path of the electrons before entering the field and after leaving the field is a straight line as there is no resultant force acting on them.

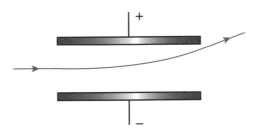

When moving between the plates:

- the electrons travel equal distances in successive equal time intervals in the direction parallel to the plates
- the electrons travel increasing distances in successive equal time intervals in the direction perpendicular to the plates.

Comparing electric and gravitational fields

AQA A	M4	EDEXCEL B	M5
AQA B	M5	OCR A	M4
EDEXCEL A	M5	WJEC	M5

There are similarities and differences between electric and gravitational fields:

- electric field strength is defined as force per unit charge, gravitational field strength is defined as force per unit mass
- electric potential and gravitational potential are defined in similar ways
- the electric field due to a point charge is similar to the gravitational field of a point mass
- electric fields can attract or repel charged objects, gravitational fields can only attract masses.

Progress check

1 Calculate the electric field strength due to a point charge of 3.0 μC at a distance of 0.10 m from the charge.
   $k = 9.0 \times 10^9$ N m^2 C^{-2}

2 The potential difference between two parallel plates is 300 V. They are placed 0.15 m apart.
   a Calculate the value of the electric field strength between the plates.
   b Calculate the size of the force on an electron, charge – 1.6 × 10^{-19} C, placed midway between the plates.
   c Explain how the size of the force on the electron varies as it moves from the negative plate to the positive plate.

c It stays the same as the field strength does not vary.
b 3.2 × 10^{-16} N
2 a 2.0 × 10^3 N C^{-1}
1 2.7 × 10^6 N C^{-1}

3.3 Capacitors

After studying this section you should be able to:

- describe the action of a capacitor and calculate the charge stored
- relate the energy stored in a capacitor to a graph of charge against voltage
- explain the significance of the time constant of a circuit that contains a capacitor and a resistor

LEARNING SUMMARY

The action of a capacitor

AQA A	M4	OCR A	M4
AQA B	M4	OCR B	M4
EDEXCEL A	M5	NICCEA	M5
EDEXCEL B	M4	WJEC	M4

Capacitors store charge and energy. They have many applications, including smoothing varying direct currents, electronic timing circuits and powering the memory to store information in calculators when they are switched off.

A capacitor consists of two parallel conducting plates separated by an insulator. When it is connected to a voltage supply charge flows onto the capacitor plates until the potential difference across them is the same as that of the supply. The charge flow and the final charge on each plate is shown in the diagram.

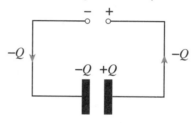

> When a capacitor is charging, charge flows in all parts of the circuit except between the plates.

As the capacitor charges:

- charge $-Q$ flows onto the plate connected to the negative terminal of the supply
- charge $-Q$ flows off the plate connected to the positive terminal of the supply, leaving it with charge $+Q$
- the capacitor plates always have the same quantity of charge, but of the opposite sign
- no charge flows between the plates of the capacitor.

Capacitance

AQA A	M4	OCR A	M4
AQA B	M4	OCR B	M4
EDEXCEL A	M5	NICCEA	M5
EDEXCEL B	M4	WJEC	M4

The capacitor shown in the diagram above is said to store charge Q, meaning that this is the amount of charge on each plate.

When a capacitor is charged, the amount of charge stored depends on:

- the voltage across the capacitor
- its capacitance: i.e. the greater the capacitance, the more charge is stored at a given voltage.

> As the capacitor plates have equal amounts of charge of the opposite sign, the total charge is actually zero.
>
> However, because the charges are separated they have energy and can do work when they are brought together.

The **capacitance** of a capacitor, C, is defined as:

$$C = \frac{Q}{V}$$

Where Q is the charge stored when the voltage across the capacitor is V. Capacitance is measured in farads (F).
1 farad is the capacitance of a capacitor that stores 1 C of charge when the p.d. across it is 1 V.

KEY POINT

One farad is a very large value of capacitance. Common values of capacitance are usually measured in picofarads (1 pF = 1.0×10^{-12} F) and microfarads (1 μF = 1.0×10^{-6} F).

Combining capacitors

AQA B	M4	NICCEA	M5
EDEXCEL A	M5	WJEC	M4

> The effect of adding capacitors in series is to reduce the capacitance. When an additional capacitor is added, there is less p.d. across each one so less charge is stored.

Like resistors, capacitors can be connected in **series** or **parallel** to achieve different values of capacitance.

When capacitors in series are connected to a voltage supply:

- no matter what the value of its capacitance, each capacitor in the combination stores the same amount of charge, since any one plate can only lose or gain the charge gained or lost by the plate that it is connected to
- the total charge stored by a series combination is the charge on each of the two outer plates and is equal to the charge stored on each individual capacitor
- because the applied potential difference is shared by the capacitors, the total charge stored is less than the charge that would be stored by any one of the capacitors connected individually to the voltage supply.

The diagram shows the charge on the plates of three capacitors connected in series.

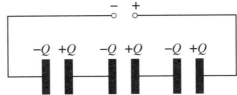

This results in the effective value of a series combination of capacitors being smaller than the lowest value capacitor in the combination.

> A common error when using this relationship is to forget to carry out the final reciprocation, giving an answer which is equal to $\frac{1}{C}$ instead of C.

> **KEY POINT**
>
> The capacitance, C, of a number of capacitors connected in series is given by the expression:
> $$\frac{1}{C} = \frac{1}{C_1} + \frac{1}{C_2} + \frac{1}{C_3}$$

In contrast to this, the effect of connecting capacitors in parallel is to increase the capacitance so that the effective value of a number of capacitors in parallel is always greater than the largest value of the combination.

When capacitors are connected in parallel:

- all the capacitors are charged to the same potential difference
- each capacitor stores the same amount of charge as it would if connected on its own to the same voltage
- adding an additional capacitor increases the total charge stored.

> The expressions for capacitors connected in series and parallel are similar to those for resistors, but the other way round.

> **KEY POINT**
>
> The capacitance, C, of a number of capacitors connected in parallel is given by the expression:
> $$C = C_1 + C_2 + C_3$$

The energy stored in a capacitor

AQA A	M4	OCR A	M4
AQA B	M4	OCR B	M4
EDEXCEL A	M5	NICCEA	M5
EDEXCEL B	M4	WJEC	M4

Energy is needed from a power supply or other source to charge a capacitor. A charged capacitor can supply the energy needed to maintain the memory in a calculator or the current in a circuit when the supply voltage is too low.

The amount of energy stored in a capacitor depends on:

- the amount of charge on the capacitor plates
- the voltage required to place this charge on the capacitor plates, i.e. the capacitance of the capacitor.

The graph below shows how the voltage across the plates of a capacitor depends on the charge stored.

When a charge ΔQ is added to a capacitor at a potential difference V, the work done is ΔQV. The total work done in charging a capacitor is $\Sigma\Delta QV$.

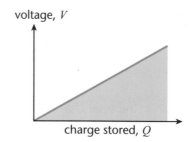

The shaded area between the graph line and the charge axis represents the energy stored in the capacitor.

The energy, E, stored in a capacitor is given by the expression
$$E = \tfrac{1}{2}QV = \tfrac{1}{2}CV^2$$
where Q is the charge stored on a capacitor of capacitance C when the voltage across it is V.

Charging and discharging a capacitor

AQA A	M4	OCR A	M4
AQA B	M4	OCR B	M4
EDEXCEL A	M6	NICCEA	M5
EDEXCEL B	M4	WJEC	M4

Having a resistor in the circuit means that extra work has to be done to charge the capacitor, as there is always an energy transfer to heat when charge flows through a resistor.

Key points from AS

- **The relationship between resistance, current and voltage.**
  Revise AS section 2.1

When a capacitor is charged by connecting it directly to a power supply, there is very little resistance in the circuit and the capacitor seems to charge instantaneously. This is because the process occurs over a very short time interval. Placing a resistor in the charging circuit slows the process down. The greater the values of resistance and capacitance, the longer it takes for the capacitor to charge.

The diagram below shows how the current changes with time when a capacitor is charging.

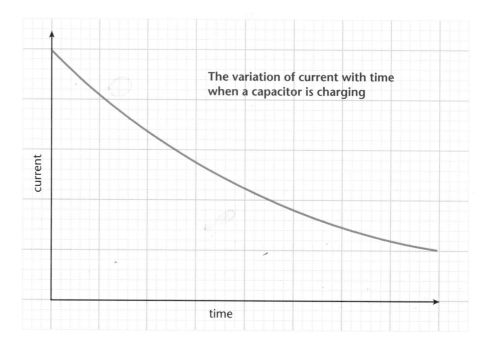

The variation of current with time when a capacitor is charging

As the p.d. across the capacitor rises, that across the resistor falls, reducing the current.

This graph shows that:

- the charging current falls as the charge on the capacitor, and the voltage across the capacitor, rise
- the charging current decreases by the same proportion in equal time intervals.

The second bullet point shows that the change in the current follows the same pattern as the activity of a radioactive isotope. This is an example of an **exponential change**, the charging current decreases exponentially.

Key points from AS

- **The relationship between current and charge flow.**
  Revise AS section 2.2

The graph shown above can be used to work out the amount of charge that flows onto the capacitor by estimating the area between the graph line and the time axis. Since *current = rate of flow of charge* it follows that:

> **KEY POINT**
>
> On a graph of current against time, the area between the graph line and the time axis represents the charge flow.

To calculate the charge flow:

- estimate the number of whole squares between the graph line and the time axis
- multiply this by the 'charge value' of each square, obtained by calculating $\Delta Q \times \Delta t$ for a single square.

The time constant

AQA A	M4	OCR A	M4
AQA B	M4	OCR B	M4
EDEXCEL A	M6	NICCEA	M5
EDEXCEL B	M4	WJEC	M4

When a capacitor is charging or discharging, the amount of charge on the capacitor changes exponentially. The graphs in the diagram show how the charge on a capacitor changes with time when it is charging and discharging.

Graphs showing the change of voltage with time are the same shape.
Since $V = \dfrac{Q}{C}$, it follows that the only difference between a charge–time graph and a voltage–time graph is the label and scale on the *y*-axis.

These graphs show the charge on the capacitor approaching a final value, zero in the case of the capacitor discharging, but never quite getting there.

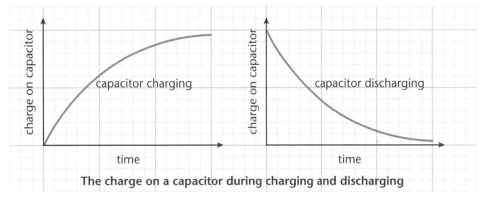

The charge on a capacitor during charging and discharging

The rate at which the charge on a capacitor changes depends on the **time constant** of the charging or discharging circuit.

> **KEY POINT**
>
> The time constant, τ, of a capacitor charge or discharge circuit is the product of the resistance and the capacitance:
> $$\tau = RC$$
> τ is measured in *s*.

The greater the values of *R* and *C* the longer the charge or discharge process takes. Knowledge of the values of *R* and *C* enables the amount of charge on a capacitor to be calculated at any time after the capacitor has started to charge or discharge. This is useful in timing circuits, where a switch is triggered once the charge, and therefore p.d., has reached a certain value.

e is a constant that has a value of 2.72. It is the base of the natural log and exponential functions.

The time constant τ represents:

- the time it takes for the charge on a capacitor to fall to $\frac{1}{e}$ of its initial value when a capacitor is discharging

- the time it takes for the charge on a capacitor to rise to $1-\frac{1}{e}$ of its final value when the capacitor is charging

The role of the time constant is similar to that of half-life in radioactive decay. When a capacitor is discharging, $\frac{1}{e^2}$ of the initial charge remains after time 2τ and $\frac{1}{e^3}$ remains after 3τ.

The exponential function e is used to calculate the charge remaining on a capacitor that is discharging.

Here e is the exponential function, the inverse of natural log, ln. Take care not to confuse this with the EXP button on a calculator, which is used for entering powers of 10.

> **KEY POINT**
>
> The charge, Q, on a capacitor of capacitance C, remaining time t after starting to discharge is given by the expression
>
> $$Q = Q_0 e^{-t/\tau}$$
>
> where Q_0 is the initial charge on the capacitor.

This expression shows that when t is equal to τ, i.e. after one time constant has elapsed, the charge remaining is equal to $Q_0 e^{-1}$, or $\frac{Q_0}{e}$

Progress check

1   The charge on a capacitor is 3.06×10^{-4} C when the p.d. across it is 6.5 V. Calculate the capacitance of the capacitor.

2   Three capacitors, of capacitance 2 μF, 3 μF and 6 μF, are connected together. Calculate the effective capacitance when they are connected:
    a   in series
    b   in parallel.

3   A 50 μF capacitor is charged to a p.d. of 360 V. Calculate the energy stored in the capacitor.

3   3.24 × 10⁻³ J
    b   11 μF
2   a   1 μF
1   4.71 × 10⁻⁵ F.

3.4 Magnetic fields

After studying this section you should be able to:

● describe the magnetic fields inside a solenoid and around a wire when current passes in them
● understand the meaning of magnetic field strength
● calculate the size and direction of the force that acts on a current in a magnetic field

Magnetic fields

Magnetic fields:

● are due to permanent magnets and electric currents
● affect permanent magnets and electric currents.

Like other fields, magnetic fields are represented by lines with arrows. The arrows show the direction of the force at any point in the field. The convention when drawing magnetic fields is that the arrows show the direction of the force that would be exerted on the N-seeking pole of a permanent magnet placed at that point.

> Although the field lines are often curves, the force at any point acts in a straight line.

The diagrams show the field patterns around a bar magnet and between two different arrangements of pairs of magnets. Where equal-sized forces act in opposite directions, the result is a neutral point. At a neutral point, the resultant magnetic force is zero.

> How many forces act on a N-seeking pole placed at the neutral point shown in the middle diagram?

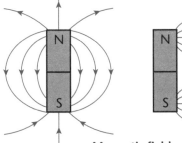
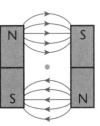
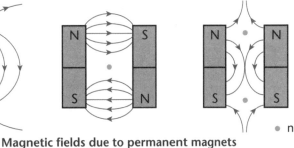

● neutral point

Magnetic fields due to permanent magnets

Magnetic field strength

A current-carrying wire placed in a magnetic field experiences a force provided that it is not parallel to the field. The force has its maximum value when the current is perpendicular to the field. Fleming's left hand rule shows the direction of the force on any current that has a component which is perpendicular to a magnetic field. The diagram illustrates Fleming's rule and its application to a simple motor.

> The force on any electric current that is parallel to a magnetic field is zero – there is no force.

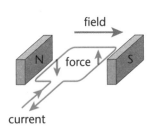

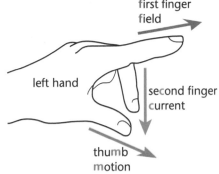

1 point your first finger in the direction of the magnetic field.

2 point your second finger in the direction of the current.

3 your thumb points in the direction of motion.

In electric and gravitational fields, the strength of the field is defined in terms of the force per unit mass or charge. There is no obvious equivalent 'unit of magnetism'.

The size of the force on a current-carrying conductor in a magnetic field depends on:

- the size of the current
- the length of conductor in the field
- the orientation of the conductor relative to the field
- the strength of the magnetic field.

> These factors are used to define the 'unit' of magnetism as *current length*, the product of the current and the length of the conductor in the field.

> **KEY POINT**
>
> These factors are used to define the strength of a magnetic field as:
>
> the force per unit (current × length) perpendicular to the field.
>
> Magnetic field strength is also known as *flux density*. Its symbol is *B* and the unit is the tesla (T).

With the current and magnetic field directions shown in the diagram below the force is into the paper.

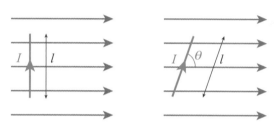

> **KEY POINT**
>
> The force on a current-carrying conductor in a magnetic field is given by the expressions:
>
> when the conductor is perpendicular to the field lines:
> $$F = BIl$$
> when the angle between the conductor and the field lines is θ,
> $$F = BIl\sin\theta$$

> You should always specify the directions of the quantities when using these relationships to define magnetic field strength.

These relationships, together with the diagrams, can be used to define magnetic field strength.

The force on a moving charge

AQA A	M4	OCR A	M4
AQA B	M5	OCR B	M5
EDEXCEL A	M5	NICCEA	M5
EDEXCEL B	M4	WJEC	M5

Any moving charge constitutes an electric current. In a vacuum tube, such as that in a television or a cathode ray oscilloscope, magnetic fields are used to exert forces on the electrons as they travel through the tube.

The size of the force on a charged particle moving in a magnetic field depends on:

- the charge on the particle
- the speed of the particle
- the strength of the magnetic field.

For a charge Q moving with velocity v, the equivalent of (current × length) is (charge × velocity) or Qv.

> Remember when using Fleming's rule that the current direction is conventional current, taken to be from + to –. The current due to an electron beam is in the opposite direction to that of the electron movement.

> **KEY POINT**
>
> A charge Q moving with velocity v perpendicular to a magnetic field experiences a force:
> $$F = BQv$$

The direction of the force is given by Fleming's left hand rule.

Fields due to currents

The diagram shows the magnetic field around a wire when a current passes in it.

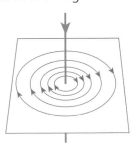

The magnetic field pattern is a set of concentric circles around the wire. The change in the spacing of the circles shows that the magnetic field strength decreases with increasing distance from the wire. The magnetic field strength depends on:

Iron has a high permeability. This explains why a coil of wire wound on an iron core has a much greater magnetic field strength than a similar coil with an air core.

- the size of the current
- the distance from the wire
- a constant called **permeability**, μ, which is a measure of the extent to which the surrounding medium reinforces the magnetic field.

The symbol H represents the henry, a unit of electromagnetic induction.

> **KEY POINT**
>
> The magnetic field strength, B, due to a current, I, passing in a wire is given by the expression:
>
> $$B = \frac{\mu_0 I}{2\pi r}$$
>
> where μ_0 is the permeability of free space and has the value $4\pi \times 10^{-7}$ H m^{-1} and r is the distance from the wire.

The magnetic field pattern due to a current passing in a **solenoid** is shown in the diagram below.

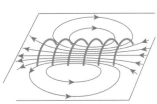

On the outside of the solenoid, the field resembles that of a bar magnet. Inside, away from the ends of the solenoid, the parallel field lines show that the magnetic field strength is uniform.

The strength of this uniform field:

- is independent of the diameter of the solenoid
- depends on the number of turns of wire per metre of length, n.

> **KEY POINT**
>
> The magnetic field strength, B, inside a solenoid is given by the expression:
>
> $$B = \mu_0 n I$$
>
> where n is the number of turns of wire per metre, μ_0 is the permeability of free space and I is the current in the solenoid.

Progress check

1 Two parallel wires each carry a current in the same direction.
   Use Fleming's rule to determine whether the force between them is attractive or repulsive.

2 The magnetic field strength between two permanent magnets is 5.6×10^{-3} T.
   A wire carrying a current of 3.5 A is placed at right angles to their magnetic field. The length of wire within the field is 0.1 m.
   Calculate the size of the force on the wire.

3 An electron, charge -1.6×10^{-19} C, moves at a velocity of 2.4×10^7 m s^{-1} perpendicular to a magnetic field of strength 4.5×10^{-2} T.
   Calculate the size of the force on the electron.

3 1.73×10^{-13} N
2 1.96×10^{-3} N
1 Attractive.

3.5 Circular orbits

LEARNING SUMMARY

After studying this section you should be able to:

- describe the forces acting on planets, moons and satellites
- explain how charged particles are accelerated in a cyclotron

Movement in the Solar System

AQA A   M4   NICCEA   M4
AQA B   M4

Mercury and Pluto have highly elliptical orbits, the other planets follow paths that are very close to being circles.

All planetary movement in the Solar System is anticlockwise, when viewed from above the North Pole. The further a **planet** is from the Sun, the slower the speed in its orbit. Although the orbits of the planets are ellipses, for most planets they are so close to circles that our understanding of circular motion can be applied.

Planets can be considered to be:

- moving at constant speed in a circle around the Sun
- accelerating towards the Sun with centripetal acceleration v^2/r.

In space there are no resistive forces since the planets move through a vacuum. The only forces acting on them are gravitational. Gravitational attraction between a planet and the Sun provides the unbalanced force required to cause the centripetal acceleration.

The diagram shows the attractive forces between the Sun and a planet.

This diagram shows that the gravitational force on a planet acts at its centre of mass and is directed towards the Sun's centre of mass.

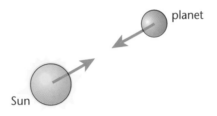

planet

Sun

The force on the planet is:

- equal in size and opposite in direction to that on the Sun
- at right angles to its direction of motion
- the unbalanced, centripetal force required to maintain circular motion.

Asteroids in the asteroid belt, between Mars and Jupiter, have a wide range of masses but similar orbit times.

By equating the gravitational force to mv^2/r, it emerges that the orbital speed depends only on the orbital radius and not on the mass of the planet.

> **KEY POINT**
>
> The centripetal force required to keep a planet in a circular orbit is the gravitational force between the planet and the Sun:
>
> $$\frac{M_p v^2}{r} = \frac{GM_s M_p}{r^2}$$
>
> so
>
> $$v^2 r = GM_s$$
>
> where M_s is the mass of the Sun and M_p is the mass of the planet.

Satellites

AQA A   M4   NICCEA   M4
AQA B   M4

The relationship also applies to the Earth's natural satellite, the Moon.

The relationship between the orbital speed and radius of a planet can be applied to the orbit of a **satellite** around the Earth by replacing the mass of the Sun, M_s, with that of the Earth, M_E. This enables the speed of a satellite to be calculated at any orbital radius.

Some communications satellites occupy geo-synchronous orbits. A satellite in a geo-synchronous orbit:

- orbits above the equator
- remains in the same position relative to the Earth's surface
- has an orbit time of 24 hours.

The radius of a geo-synchronous orbit can be calculated from $v^2r = GM_E$. As there are two unknowns in this equation, v can be written as $2\pi r/t$ to work out the value of r.

Circular orbits in magnetic fields

AQA A   M4   EDEXCEL B   M4
AQA B   M4   OCR A   M4

When a charged particle moves at right angles to a magnetic field, the magnetic force on the particle is perpendicular to both its direction of motion and the magnetic field. This can result in circular motion.

The diagram shows the path and the force on an electron moving in a magnetic field directed into the paper.

When applying Fleming's rule to electrons, remember that the direction of the current is opposite to that of the electrons' motion.

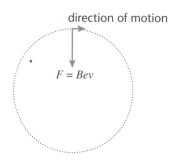

direction of motion

$F = Bev$

The electron follows a circular path, the magnetic force being the unbalanced force required to cause acceleration towards the centre of the circle. The radius of the circular path is proportional to the speed of the electron.

For an electron, $Q = e$, so the relationship is

$Be = \dfrac{mv}{r}$.

When a charge Q moves in a circular path in a magnetic field of strength B:
$$BQv = \frac{mv^2}{r}$$
so $$BQ = \frac{mv}{r}$$

KEY POINT

The cyclotron

AQA A   M4   EDEXCEL A   M6
AQA B   M4   NICCEA   M6

A **cyclotron** uses a magnetic field to force charged particles to move in a circular path, and an electric field to accelerate them as they travel around the circle. As the charged particles accelerate, the increase in speed results in an increase in the radius of the circle, so they spiral outwards.

A cyclotron consists of two D-shaped halves called dees. A magnetic field acting at right angles to the plane of the dees causes a beam of charged particles to follow a circular path. Particles such as protons and alpha particles are both suitable for use in cyclotrons.

Particles accelerated in a cyclotron are used to probe atomic nuclei and for treating some cancers.

The diagram shows the path of protons produced at the centre of the cyclotron.

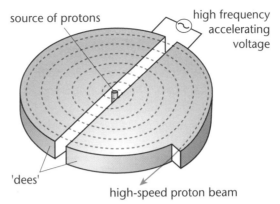

source of protons

high frequency
accelerating
voltage

'dees'

high-speed proton beam

In a cyclotron:

- the beam of charged particles is accelerated as it passes from one dee into the other
- this occurs because of the alternating electric field which changes polarity so that it attracts the particles as they enter a dee
- the frequency of the alternating voltage must be equal to the frequency of rotation of the particles
- the radius of orbit increases as the particles accelerate.

The frequency of rotation of the charged particles in a cyclotron matches that of the accelerating voltage and does not depend on the speed of the particles. The value of the magnetic field strength can be adjusted to achieve the desired frequency.

> If the frequency of the accelerating voltage is fixed, each orbit takes the same time. With an increase in the radius of successive orbits, the particles travel increasing distances in a given time period.

> **KEY POINT**
>
> The frequency of rotation of a charged particle in a cyclotron, f, is related to the magnetic field strength, B, by the expression:
>
> $$f = \frac{BQ}{2\pi m}$$
>
> where Q is the charge on a particle of mass m.

Progress check

1 The Moon orbits the Earth, mass 6.0×10^{24} kg, at a radius of 3.84×10^8 m.
   $G = 6.7 \times 10^{-11}$ N m² kg⁻².
   Calculate:
   a the speed of the Moon in its orbit.
   b the time it takes for the Moon to complete one orbit.

2 An electron of mass $m_e = 9.10 \times 10^{-31}$ kg and charge, $e = 1.60 \times 10^{-19}$ C travels at a speed of 2.10×10^7 m s⁻¹ in a circular orbit at right angles to a magnetic field.
   The magnetic field strength, $B = 6.5 \times 10^{-6}$ T.
   Calculate the radius of the electron orbit.

3 In the diagram of the cyclotron what is the direction of the magnetic field?

3 Upwards.
2 18.4 m
b 2.36×10^6 s
1 a 1.02×10^3 m s⁻¹

3.6 Electromagnetic induction

After studying this section you should be able to:

- *calculate the flux linkage through a coil of wire in a magnetic field*
- *explain how electromagnetic induction occurs due to changes in flux linkage*
- *apply Faraday's law and Lenz's law*

LEARNING SUMMARY

Flux and flux linkage

AQA A	M4	OCR A	M4
AQA B	M5	OCR B	M5
EDEXCEL A	M5	NICCEA	M5
EDEXCEL B	M4	WJEC	M5

Flux provides a useful model for explaining the effects of magnetic fields.

The current view is that these forces can be attributed to 'exchange particles'.

Almost everything we do, apart from sleeping in the dark, relies on **electromagnetic induction**. Induction is used to generate electricity in power stations and to transform its voltage as it passes through the distribution system.

The effects of induction are explained by using the concept of **flux**. Although the existence of flux has long been discredited, an awareness of its meaning is useful to understand the laws of induction as set out by Faraday and Lenz.

Like gravitational and electric fields, magnetic fields act at a distance. Magnetic field patterns are used to show the forces that are exerted around a magnet or electric current. These forces are exerted without any physical link between the magnet or current that causes the field and a magnetic material or current placed within the field. In the days of Faraday and Lenz, they were attributed to the effects of flux.

When drawing magnetic field patterns:

- the relative strength at different points in the field is shown by the separation of the field lines
- the closer the lines are together, the stronger the field
- these field line represent **magnetic flux**, which is imagined as occupying the space around a magnet and being responsible for the effect of a magnet field.

To integrate the flux model with today's explanation of magnetic effects in terms of magnetic field strength, this can be thought of in terms of a flux density, being represented by the concentration of magnetic field lines. Flux density is the flux per unit area so flux is now defined in terms of the magnetic field strength and the area that the field permeates.

This definition relates the equivalence of the modern concept of magnetic field strength to that of the older 'flux density' concept.

> The magnetic flux, ϕ, through an area, A, is defined as the product of the magnetic field strength and the area normal to the field.
>
> $$\Phi = B \times A$$
>
> Magnetic flux is measured in webers (Wb) where 1 Wb is the flux through an area of 1 m² normal to a uniform field of strength 1 T.

KEY POINT

The diagram shows the flux through a rectangular coil in a uniform magnetic field.

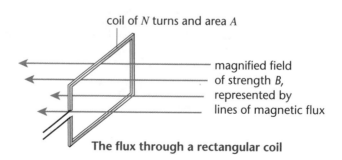

coil of N turns and area A

magnified field of strength B, represented by lines of magnetic flux

The flux through a rectangular coil

Movement of the coil parallel to the field does not induce an e.m.f., since no field lines are being 'cut'.

The induced e.m.f. has its greatest value when the movement of the coil is perpendicular to the field.

When the coil is rotated, it 'cuts' through the flux, or field lines and an e.m.f. is induced.

The size, or magnitude, of the induced e.m.f. depends on:

- the amount of flux through the coil
- the speed of rotation
- the number of turns on the coil.

Each turn on the coil has a flux linkage which changes as the coil rotates.

The flux linkage of a coil of N turns is $N\Phi$, where Φ is the flux through the coil.

Faraday's law

AQA A	M4	OCR A	M4
AQA B	M5	OCR B	M5
EDEXCEL A	M5	NICCEA	M5
EDEXCEL B	M4	WJEC	M5

In a power station, electricity is generated by an electromagnet spinning inside copper coils.

Electromagnetic induction occurs whenever the magnetic field through a conductor changes. This can be due to a conductor moving through a magnetic field or a conductor being in a fixed position within a changing magnetic field, such as that due to an alternating current. Both of these result in an e.m.f. being induced in the conductor.

Examples of electromagnetic induction include:

- moving a magnet inside a wire coil
- generating the high voltage necessary to ionise the vapour in a fluorescent tube and cause the spark needed to ignite the explosive mixture in a petrol engine
- changing the voltage of an alternating current, using a transformer.

The diagram below shows the difference in the size of the e.m.f. when a magnet is moved at different speeds in a coil.

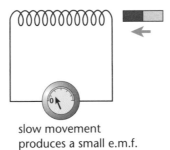
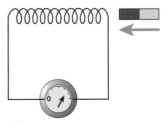

slow movement
produces a small e.m.f.

faster movement
produces a bigger e.m.f.

Faraday's law relates the size of the induced e.m.f. to the change in flux linkage.

To generate the high voltage needed to cause a spark, the flux has to change rapidly. This happens when the current in an electromagnet is switched off.

> **KEY POINT**
>
> Faraday's law states that:
>
> *the size of the induced e.m.f. is proportional to the rate of change of flux linkage.*
>
> As the proportionality constant is equal to 1, for a uniform rate of change of flux linkage this can be written as:
>
> $$\text{magnitude of induced e.m.f.} = N\frac{\Delta\Phi}{\Delta t}$$
>
> where $\Delta\Phi$ is the change of flux in time Δt.

What direction?

AQA A	M4	OCR A	M4
AQA B	M5	OCR B	M5
EDEXCEL A	M5	NICCEA	M5
EDEXCEL B	M4	WJEC	M5

Faraday's law can be used to work out the size of an induced e.m.f. such as that across the wingtips of an aircraft flying in the Earth's magnetic field. In Britain the Earth's field makes an angle of 20° with the vertical, see the following diagram.

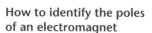

An aircraft flying in a North–South direction is cutting across the vertical component only, while flying East–West involves cutting across the horizontal component in addition.

> Unlike that of a bar magnet, the Earth's magnetic field is from South to North. It can be considered to have two components, vertical and horizontal.

The induced e.m.f. arises as a consequence of the force on the free electrons in the metal of the aircraft frame. As the aircraft travels through the air, the movement of these electrons forms a current in the opposite direction to that of flight. Fleming's left hand rule can be used to work out the direction of the force on the electrons and hence the direction of the induced e.m.f.

> All the charged particles experience a force due to their movement through a magnetic field, but the force is too weak to affect anything other than the free electrons.

In the case of an aircraft flying from North to South:

- the current is South–North
- the magnetic field being 'cut' is vertically downwards
- the force on the free electrons is towards the East.

This results in a charge imbalance and a voltage across the wingtips.

The direction of the e.m.f. induced in the aircraft and when a magnet moves into a coil of wire can be worked out using **Lenz's law**.

> If the induced e.m.f. in the aircraft caused electrons to flow from West to East, it would produce a force in a Northerly direction – opposite to the motion of the aircraft.
>
> This does not happen because there is no complete circuit.

KEY POINT

Lenz's law states that:

the direction of an induced e.m.f. is always in opposition to the change that causes it.

The diagram below shows that when the North pole of a magnet is moved into one end of a coil, the induced e.m.f. causes an induced current in an anticlockwise direction. When current passes in a coil, the magnetic field is similar to that of a bar magnet, the North pole being the end where the current passes anticlockwise.

> The direction of the induced current is reversed by reversing the magnet or its direction of movement.

The induced current when a magnet enters a coil of wire

How to identify the poles of an electromagnet

> If the induced current was in the opposite direction, it would attract the magnet into the coil and generate electricity with no energy input.

Lenz's law is a re-statement of the principle of conservation of energy; the induced current opposes the motion of the magnet so work has to be done to move the magnet against the induced magnetic field. This work is the energy transfer to the circuit needed to cause a current.

Combining Faraday's and Lenz's laws gives the equation for induced e.m.f.:

KEY POINT

$$\varepsilon = -N\frac{\Delta\Phi}{\Delta t}$$

Where ε is the induced e.m.f. The negative sign shows that the induced e.m.f. is in opposition to the change of flux causing it.

The transformer

AQA B   M5   OCR B   M5
EDEXCEL A   M5   NICCEA   M5
EDEXCEL B   M4

Transformers use changing magnetic fields to change the size of an alternating voltage. An alternating current passing in one coil (the primary) induces an e.m.f. in an adjacent coil (the secondary).

The diagram below shows the flux when the two coils are wound on an iron core.

Flux linking two transformer coils

The e.m.f. is induced whether or not there is a secondary circuit. If there is a complete circuit, there is also an induced current.

In a transformer:

- alternating current in the primary produces an alternating magnetic field
- this is reinforced by the high-permeability iron core
- the flux concentrates in the iron
- an e.m.f. is induced in the secondary because of the changing flux linkage.

Iron is easily magnetised; its magnetic domains contribute to the strength of the magnetic field.

It follows from the last bullet point that the induced e.m.f. is proportional to the number of turns on the secondary coil.

A transformer constructed from low-resistance coils on a laminated iron core is close to ideal.

> **KEY POINT**
>
> The relationship between the voltages and numbers of turns for an ideal transformer is:
>
> $$\frac{V_p}{V_s} = \frac{N_p}{N_s}$$

This states that the voltages are in the same ratio as the numbers of turns. In an ideal transformer there is no energy loss in the wires or the core so the power output from the secondary is equal to the power input to the primary and the currents are in the inverse ratio to the voltages.

Progress check

1   A rectangular coil has 25 turns and an area of 2.5×10^{-4} m^2.
    It is placed in a magnetic field of strength 6.8×10^{-6} T.
    Calculate the flux linkage when the plane of the coil is
    a   parallel to the magnetic field
    b   perpendicular to the magnetic field.

2   An aircraft flies from East to West.
    In what direction is the induced e.m.f. due to the 'cutting' of the Earth's horizontal magnetic field?

3   The flux linking a coil of 60 turns changes at the rate of 4.0×10^{-3} Wb s^{-1}.
    Calculate the size of the induced e.m.f.

3   0.24 V
2   The top of the aircraft is positive and the bottom negative
1   a   0
    b 4.25×10^{-8} Wb

Sample question and model answer

The diagram shows a capacitor connected in series with a 12.0 V power supply and a 500 Ω resistor.

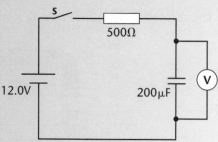

The switch is closed.

(a) What is the initial voltage across:
 (i) the capacitor? [1]

> At the instant that the switch is closed there is no charge on the capacitor, so there is no voltage across it. All the voltage is therefore across the resistor.

   0 V   1 mark

 (ii) the resistor? [1]

   12 V   1 mark

(b) Calculate the value of the current in the circuit immediately after the switch is closed. [3]

> The value of the resistor determines the initial current in the circuit.

   $I = V \div R$ 1 mark
   $= 12.0 \text{ V} \div 500 \text{ Ω}$ 1 mark
   $= 0.024 \text{ A}$ 1 mark

(c) When the capacitor is fully charged, calculate:
 (i) the charge stored on the capacitor [3]

   $Q = C \times V$ 1 mark
   $= 200 \text{ μF} \times 12.0 \text{ V}$ 1 mark
   $= 2.4 \times 10^{-3} \text{ C}$ 1 mark

 (ii) the energy stored by the capacitor. [3]

> Alternatively, ½QV should give the same answer, provided that the charge, Q, has been calculated correctly.

   $E = \frac{1}{2} C \times V^2$ 1 mark
   $= \frac{1}{2} \times 200 \text{ μF} \times 12.0^2$ 1 mark
   $= 1.44 \times 10^{-2} \text{ J}$ 1 mark

(d) The capacitor is discharged and a second 200 μF capacitor is connected in parallel with the capacitor shown in the diagram. The switch is then closed. Without doing any further calculations, explain how this affects:
 (i) the final charge stored [2]

   This is doubled (1 mark) since the potential difference is unchanged and the capacitance is doubled (1 mark).

 (ii) the final energy stored. [2]

> As the question states that further calculations are not required, give an answer based on the relationships used to calculate these quantities, with the reasons.

   This is also doubled (1 mark) as there is double the capacitance (and double the amount of charge stored) for the same potential difference (1 mark).

Practice examination questions

Throughout this section, use the following values of constants:
electronic charge, $e = -1.6 \times 10^{-19}$ C
universal gravitational constant, $G = 6.7 \times 10^{-11}$ N m² kg⁻²
$k = 1/4\pi\varepsilon_0 = 9.0 \times 10^9$ N m² C⁻²
gravitational field strength at the surface of the Earth, $g = 10.0$ N kg⁻¹

1

A hydrogen atom consists of a proton and an electron at an average separation of 5.2×10^{-11} m.

(a) Calculate the size of the force between them. [3]

(b) (i) Calculate the electric field strength that each experiences due to the other. [3]

   (ii) What is the direction of the electric field between the proton and the electron? [1]

(c) Assuming that the electron orbits the proton, calculate the speed of its orbit.
   The mass of an electron, $m_e = 9.1 \times 10^{-31}$ kg. [3]

2

The diagram shows a pair of parallel plates connected to 2500 V supply.

An oil drop of mass 0.050 g between the plates carries a charge of 10e.

(a) Calculate:

   (i) the voltage between C and D [1]

   (ii) the energy transfer when the drop moves from C to D. [3]

(b) Calculate the energy transfer when the drop moves from A to B. [1]

(c) The voltage is changed so that the drop is stationary between the plates.

   (i) What must the polarity of the plates be to achieve this? [1]

   (ii) Calculate the voltage required. [3]

3

Use this data to answer the questions:
The mass of the Earth = 6.0×10^{24} kg.
The mass of the Moon = 7.4×10^{21} kg.
The distance between the Earth and the Moon = 3.8×10^8 m.

(a) (i) Calculate the strength of the Earth's gravitational field at the Moon's orbit. [3]

   (ii) Calculate the size of the Earth's pull on the Moon. [2]

   (iii) Use the answer to (ii) to calculate the Moon's period of rotation around the Earth.

Practice examination questions (continued)

(b) Tides are due to the combined effect of the Moon and the Sun.

(i) Calculate the size of the Moon's pull on 1.0 kg of water. [3]

(ii) Suggest why the pull of the Sun has much less effect than the pull of the Moon on tides. [3]

(iii) The diagram shows two positions of the Moon relative to the Earth and the Sun.

Suggest why tides are higher when the Moon is in position A than when it is in position B. [2]

4

A moving coil loudspeaker consists of a cylindrical permanent magnet and an electromagnet. The electromagnet is positioned between the poles of the fixed magnet. The diagram shows the arrangement.

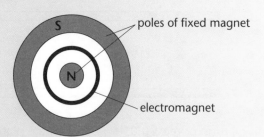

(a) What is the direction of the magnetic field between the poles of the fixed magnet? [1]

(b) The current in the electromagnet passes in a clockwise direction.
What is the direction of the force on the electromagnet? [1]

(c) Explain why the electromagnet vibrates when an alternating current passes in it. [2]

(d) The value of the magnetic field strength due to the fixed magnet is 0.85 T at the position of the electromagnet.
The electromagnet consists of a coil of wire of 150 turns and diameter 5.0 cm.
Calculate the size of the force on the coil when a current of 0.055 A passes in it. [3]

5   Electrons can be made to move in a circular path by firing them into a region where there is a magnetic field acting at right angles to their velocity.

The diagram shows a device for producing high-speed electrons.

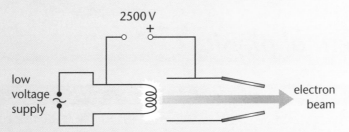

(a) Calculate:

   (i)   the kinetic energy of an electron in the accelerated beam   [3]

   (ii) the speed of an electron in the accelerated beam.
          The mass of an electron, m_e = 9.1 × 10⁻³¹ kg.   [3]

(b) The electron beam passes into a magnetic field directed into the paper.

   (i)   What is the direction of the force on the electron beam as it enters the
          magnetic field?   [1]

   (ii) Explain why the electrons follow a circular path.   [2]

   (iii) The strength of the magnetic field is 2.0 mT.
          Calculate the radius of the circular path of the electrons.   [3]

6

A rectangular coil of wire is placed so that its plane is perpendicular to a magnetic field of strength 0.15 T. This is shown in the diagram.

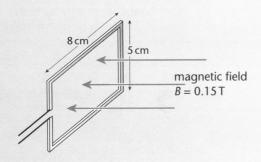

(a) Calculate the magnetic flux through the coil:

   (i)   when the coil is in the position shown in the diagram   [3]

   (ii) when the coil is turned through 90°.   [1]

(b) The coil is connected to a high-resistance voltmeter.

   (i)   Describe and explain how the reading on the voltmeter changes as the coil
          is rotated through 360°.   [4]

   (ii) What device makes use of this effect?   [1]

(c) State three ways of increasing the size of the induced voltage.   [3]

Chapter 4
Particle physics

The following topics are covered in this chapter:

- Thermal physics
- Radioactive decay
- Energy from the nucleus
- Probing matter

4.1 Thermal physics

After studying this section you should be able to:

- explain how temperature is related to internal energy
- calculate the energy transfer required for a given temperature change
- recall and use the ideal gas equation

LEARNING SUMMARY

Energy of individual particles

AQA A	AS	OCR A	M4
AQA B	M4	OCR B	M4
EDEXCEL A	AS	NICCEA	M4
EDEXCEL B	M5	WJEC	M4

All materials are made up of particles, and particles have energy. The particles of an ideal gas have only kinetic energy since forces between the particles are negligible. However, this cannot be true of a solid or a liquid or the particles would not stay in that phase, and the particles in real gases exert both attractive and repulsive forces on each other during collisions.

Imagine a collision between two particles in a real gas.

- As the particles approach, the attractive forces cause an increase in the speed and kinetic energy.

> The speed of colliding particles is momentarily zero as their direction of travel is reversed.

- The particles then get closer and the speed decreases, being momentarily zero before they start to separate.

- The process is then reversed.

There is an interchange between potential (stored) energy and kinetic energy during the collision, as shown in the diagram.

> As the separation between two particles decreases, the net force changes from attractive to repulsive. The equilibrium position is where these forces are equal in size, so the resultant force is zero.

particles speed up due to
short-range attractive forces

particles slow down due to
close-range repulsive forces

The particles of real gases, like those of solids and liquids, have both potential and kinetic energy. The distribution of energy between kinetic and potential is constantly changing and so is said to be random.

The total amount of energy of the particles in an object is known as its internal energy. At a constant temperature the internal energy of an object remains unchanged, but the contributions of individual particles to that energy change due to the transfer of energy during interactions between particles.

Measuring temperature

| EDEXCEL B | AS | NICCEA | M4 |
| OCR A | M4 | WJEC | M4 |

Suitable thermometric properties include volume, resistance and length.

The Celsius scale, based on the ice-point and the steam-point, is a useful scale for everyday measurements of temperature. However, there are some drawbacks:

- thermometers which use different thermometric properties only agree at the points of calibration
- the Celsius scale is not an absolute scale, so 20°C is not twice as hot as 10°C.

Since temperature is related to the internal energy of a material, on an absolute scale of temperature:

- the zero should correspond to the minimum internal energy of the particles
- there should be a unique value for any temperature, which does not depend on the instrument being used to measure it
- 20 units should be twice as hot as 10 units.

The thermodynamic scale of temperature, based on the behaviour of an ideal gas, satisfies these conditions.

Note that the kelvin is not called a degree and does not have a ° in the unit.

> **KEY POINT**
>
> The absolute scale of temperature is called the **kelvin** scale; its unit is the kelvin (K) and its relationship to the Celsius scale is:
>
> $$T/K = \theta/°C + 273$$

The diagram shows the relationship between Celsius and kelvin temperatures.

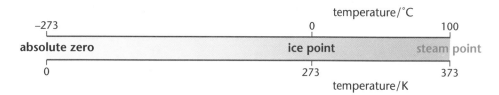

Being specific

AQA A	AS	OCR A	M4
AQA B	M4	OCR B	M4
EDEXCEL A	AS	WJEC	M4

Changing the temperature of an object involves a change in its internal energy, the total potential and kinetic energy of the particles.

The energy transfer required to change the temperature of an object depends on:

- the temperature change
- the mass of the object
- the material the object is made from.

These are all taken into consideration in the concept of specific heat capacity. The term 'specific' means 'for each kilogram' so any physical measurement that is described as 'specific' refers to 'per kilogram of material'.

> **KEY POINT**
>
> The specific heat capacity of a material, c, is defined as:
> the energy transfer required to change the temperature of 1 kg of the material by 1°C.
>
> $$\Delta E = mc\,\Delta\theta$$
>
> where c, specific heat capacity, is measured in J kg^{-1} °C^{-1} or J kg^{-1} K^{-1}

When using this relationship the temperatures can be in either Celsius or kelvin since the temperature change is the same in each case.

Changing phase

AQA A   AS   OCR B   M4
OCR A   M4

If you try to squash a solid or a liquid, the particles are pushed closer together and they repel each other. Stretching has the opposite effect; when the separation of the particles is increased the forces are attractive. At increased separations the particles have increased potential energy and this energy has to be supplied for any process that involves expansion to take place.

> The term phase means whether the substance is a solid, liquid or gas.

When a substance changes phase there is a change in the potential energy of the particles.

For most substances there is a small increase in potential energy when changing from solid to liquid and a much bigger change from liquid to gas.

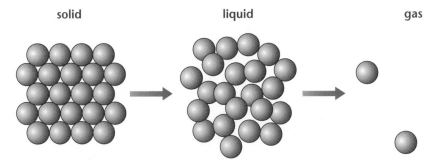

solid   **liquid**   **gas**

> The increase in particle separation during a change of phase from solid to liquid is small, but that during a change of phase from liquid to gas is large.

> On each horizontal part of the curve the substance exists in two phases. Can you identify them?

The diagram below shows a typical temperature–time graph as a solid is heated at a constant rate and passes through the three phases. Where the curve is horizontal it shows that energy is being absorbed with no change in temperature. This corresponds to a change of phase. Note that much more energy is absorbed during the change from liquid to gas than during the change from solid to liquid.

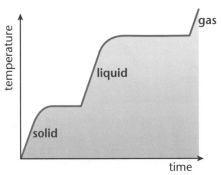

An ideal gas

AQA A   AS   OCR A   M4
AQA B   M4   OCR B   M4
EDEXCEL A   AS   NICCEA   M4
EDEXCEL B   M5   WJEC   M4

The concept of an ideal gas is useful because the behaviour of all gases is close to ideal provided that the gas is neither at a high pressure nor close to its boiling point. The ideal gas equation gives the relationship between the three variables pressure, volume and temperature.

For an ideal gas these variables are related by the single equation:

$$pV/T = \text{constant}$$

> 1 mole of gas consists of the Avagadro constant, N_A, of particles. The value of the Avagadro constant is 6.02×10^{23} mol^{-1}.

The value of the constant depends only on the amount of gas. For 1 mole of any gas under ideal conditions it has the value of 8.3 J mol^{-1} K^{-1}. This constant is called the molar gas constant and has the symbol R. Doubling the number of particles by considering two moles of gas has the effect of doubling the value of the constant, so:

> **KEY POINT**
>
> The equation of state for an ideal gas is:
> $$pV = nRT$$
> where n is the number of moles of gas and R is the molar gas constant.

In an ideal gas, all the internal energy exists as kinetic energy of the particles. The internal energy is proportional to thermodynamic temperature.

> **KEY POINT**
>
> The mean kinetic energy of the particles of an ideal gas is proportional to the kelvin temperature.

This means that doubling the thermodynamic temperature of a gas results in the mean (speed)^2 being doubled, so the mean speed is increased by a factor of $\sqrt{2}$.

Progress check

1 1.5 kg water at 10°C is placed in a kettle. Calculate the energy transferred to the water in bringing it to the boil. The specific heat capacity of water, $c = 4.2 \times 10^3$ J kg^{-1} °C^{-1}.

2 Calculate the molar volume of an ideal gas at atmospheric pressure $(1.01 \times 10^5$ Pa) and a temperature of 0°C.

3 The mean kinetic energy of the particles of an ideal gas is 6.07×10^{-21} J at a temperature of 20°C. At what temperature is their mean kinetic energy equal to 12.14×10^{-21} J?

3 586 K or 313°C
2 2.24×10^{-2} m^3
1 5.67×10^5 J

4.2 Radioactive decay

After studying this section you should be able to:

- describe the main types of radioactive emission
- explain the effects of radioactive decay on the nucleus
- calculate the half-life of a radioactive isotope from an activity–time graph

LEARNING SUMMARY

Radiation all around us

AQA A	M5	EDEXCEL B	AS
AQA B	AS	WJEC	M5
EDEXCEL A	AS		

The term 'background radiation' is also used to describe the microwave radiation left over from 'Big Bang'. This is a different type of radiation to nuclear radiation.

Radioactive decay occurs when an atomic nucleus changes to a more stable form. This is a random event that cannot be predicted. The emissions from these nuclei are collectively called radioactivity or radiation.

We are subjected to a constant stream of radiation called background radiation. Most of this is 'natural' in the sense that it is not caused by the activities of people living on Earth. Sources of background radiation include:

- the air that we breathe; radioactive radon gas from rocks can concentrate in buildings
- the ground and buildings; all rocks contain radioactive isotopes
- the food that we eat; the food chain starts with photosynthesis. Radioactivity enters the food chain in the form of carbon-14, an unstable form of carbon that is continually being formed in the atmosphere
- radiation from space, called cosmic radiation
- medical and industrial uses of radioactive materials.

The emissions

AQA A	M5	OCR A	M4
AQA B	AS	OCR B	M5
EDEXCEL A	AS	NICCEA	M5
EDEXCEL B	AS	WJEC	M5

The range of a beta-plus particle is effectively zero, since it is annihilated when it collides with an electron.

Alpha radiation is the most intensely ionising and can cause a lot of damage to body tissue.

Although it cannot penetrate the skin, alpha emitters can enter the lungs during breathing.

Radioactive emissions are detected by their ability to cause ionisation, creating charged particles from neutral atoms and molecules by removing outer electrons. This results in a transfer of energy from the emitted particle, which is effectively absorbed when all its energy has been lost in this way. The four main emissions are alpha (α), beta-plus (β^+), beta-minus (β^-) and gamma (γ). Of these, alpha radiation is the most intensely ionising and has the shortest range, with the exception of beta-plus, while gamma radiation is the least intensely ionising and has the longest range.

- In alpha emission the nucleus emits a particle consisting of two protons and two neutrons. This has the same make-up as a helium nucleus.
- Beta-minus emission occurs when a neutron decays into a proton, emitting an electron in the process.
- In beta-plus emission a proton changes to a neutron by emitting a positron, an anti-electron.
- Gamma emission is short wavelength electromagnetic radiation.

radioactive emission	nature	charge/e	symbol	penetration	causes ionisation	affected by electric and magnetic fields
alpha	two neutrons and two protons	+2	^4_2He or $^4_2\alpha$	absorbed by paper or a few cm of air	intensely	yes
beta-minus	high-energy electron	−1	$^0_{-1}\text{e}$ or $^0_{-1}\beta$	absorbed by 3 mm of aluminium	weakly	yes
beta-plus	positron (antielectron)	+1	$^0_{+1}\text{e}$ or $^0_{+1}\beta$	annihilated by an electron		yes
gamma	short-wavelength electromagnetic radiation	none	$^0_0\gamma$	reduced by several cm of lead	very weakly	no

Notice that the electron emitted in beta-minus decay and the positron emitted in beta-plus decay have been allocated the atomic numbers –1 and +1. This is because of the effect on the nucleus when these particles are emitted.

Other types of nuclear decay include:

- electron capture, where a proton in the nucleus captures an orbiting electron and becomes a neutron. The energy lost by the electron is emitted as an X-ray
- nucleon emission; some artificially-produced isotopes decay by emitting a proton or a neutron.

Alpha and beta radiations are the most intensely ionising but they are readily absorbed and so easy to shield. Gamma radiation is very penetrating. Although it only reacts weakly with matter, its effects can be devastating. One of the safest ways of protecting against danger from gamma radiation is to maximise the distance from the source to any people. The radiation from a gamma source is emitted equally in all directions, so the intensity decreases as an inverse square law.

> In this context, intensity means the number of photons detected per square metre.

> **KEY POINT**
>
> The intensity, I, of gamma radiation detected from a source is related to the intensity of the source, I_0, by the equation:
>
> $$I = \frac{kI_0}{x^2}$$
>
> where $k = 1/4\pi$ and x is the distance from the source.

Balanced equations

AQA A	M5	OCR B	M5
EDEXCEL A	AS	NICCEA	M5
EDEXCEL B	M4	WJEC	M5
OCR A	M4		

When a nucleus decays by alpha or beta emission, the numbers of protons and neutrons are changed. Gamma emission does not change the make-up of the nucleus, but corresponds to the nucleus losing excess energy. Gamma emission often occurs alongside alpha and beta emissions, though some artificial radioactive isotopes emit gamma radiation only.

> Technetium-99 is an artificial isotope that emits gamma radiation only. It is used as a tracer in medicine. The gamma radiation can be detected outside the body and there is less risk of cell damage than with an isotope that also emits alpha or beta radiation.

The changes that take place due to alpha and beta emissions are:

- alpha; the number of protons decreases by 2 and the number of neutrons also decreases by two
- beta-minus; the number of neutrons decreases by one and the number of protons increases by one
- beta-plus; the number of neutrons increases by one and the number of protons decreases by one.

In writing equations that describe nuclear decay, both charge (represented by the atomic number, Z) and the number of nucleons (represented by the mass number, A) are conserved. The table summarises these changes and gives examples of each type of decay.

> Check that the equations given as examples are balanced in terms of charge and number of nucleons.

particle emitted	effect on A	effect on Z	example
alpha	–4	–2	$^{226}_{88}\text{Ra} \rightarrow ^{222}_{86}\text{Rn} + ^{4}_{2}\text{He}$
beta-minus	unchanged	+1	$^{14}_{6}\text{C} \rightarrow ^{14}_{7}\text{N} + ^{0}_{-1}\text{e}$
beta-plus	unchanged	–1	$^{11}_{6}\text{C} \rightarrow ^{11}_{5}\text{B} + ^{0}_{+1}\text{e}$

Rate of decay and half-life

AQA A	M5	OCR A	M4
AQA B	M5	OCR B	M4
EDEXCEL A	AS	NICCEA	M5
EDEXCEL B	M5	WJEC	M5

> **KEY POINT**
>
> The activity, or rate of decay, of a sample of radioactive material is measured in **becquerel** (Bq). An activity of 1 Bq represents a rate of decay of 1 s^{-1}.

Radioactive decay is a random process and the decay of an individual nucleus cannot be predicted. However, given a sample containing large numbers of undecayed nuclei, then statistically the rate of decay should be proportional to the number of undecayed nuclei present. Double the size of the sample and, on the average, the rate of decay should also double.

There are only two factors that determine the rate of decay of a sample of radioactive material. They are:

- the radioactive isotope involved
- the number of undecayed nuclei.

Unlike chemical reactions, radioactive decay is not affected by changes in temperature.

The relationship between the rate of decay, or activity, of a radioactive isotope and the number of undisclosed nuclei is:

$$\text{activity, } A = \lambda N$$

Where the activity is measured in becquerel, N is the number of undecayed nuclei present and λ is the **decay constant** of the substance. λ has units of s^{-1}.

When the activity of a radioactive isotope (after deducting the average background count) is plotted against time, the result is a curve that shows the activity decreasing as the number of undecayed nuclei decreases. A decay curve is shown in the diagram.

The graph shown is a plot of activity against time. A plot of number of undecayed nuclei against time would be identical but with a different scale on the y-axis.

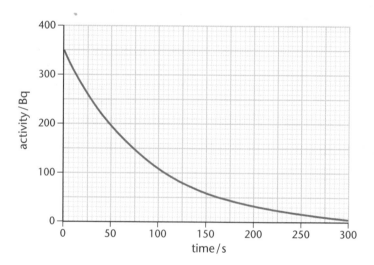

The shape of the curve is the same for all radioactive substances, but the activities and time scales depend on the size of the sample and its decay constant.

The negative sign is needed here because the *rate of change of undecayed nuclei* is always negative due to the decrease in the numbers of undecayed nuclei with increasing time.

An alternative way of expressing activity is *the rate of change of undecayed nuclei with time:*

$$\frac{\Delta N}{\Delta t} = -\lambda N$$

this leads to the relationship between the number of undecayed nuclei, N, and the number at time $t = 0$, N_0:

$$N = N_0 e^{-\lambda t}$$

The curve on the graph above shows exponential decay. In an exponential decay curve:

- the rate of change of y, represented by the gradient of the curve, is proportional to y
- in equal intervals of x the value of y always changes in the same ratio.

This second point means that it always takes the same time interval for the activity to decrease to a given fraction of any particular value. The time for the activity to halve is a measure of the **half-life** of the substance.

Note that although half-life is defined in terms of the number of undecayed nuclei, it is usually measured as the average time for the activity of a sample to halve.

> **KEY POINT**
>
> The half-life of a radioactive isotope, $t_{1/2}$ is the average time taken for the number of undecayed nuclei of the isotope to halve.

Because radioactive decay is a random process, the results of any experiment to measure activity do not fit the curve exactly and there is some variation in the change of activity when identical samples of the same material are compared. This is why the term *average* is used in the definition of half-life.

The values of half-lives range from tiny fractions of a second to many millions of years. Starting with N nuclei of a particular isotope, the number remaining after one half-life has elapsed is $N/2$, after two half-lives $N/4$ and after n half-lives it is $N/2^n$.

There is a relationship between the half-life and the decay constant of any particular radioactive isotope. The shorter the half-life, the greater the rate of decay and the decay constant.

When using this relationship, λ and $t_{1/2}$ must be measured in consistent units, e.g. λ in s^{-1} and $t_{1/2}$ in s.

> **KEY POINT**
>
> The half-life of a radioactive isotope and its decay constant are related by the equation:
> $$\lambda t_{1/2} = \ln 2 = 0.69$$

Stable and unstable nuclei

AQA A	M5	OCR B	M5
EDEXCEL A	AS	NICCEA	M5
OCR A	M4	WJEC	M5

Remember, isotopes of an element all have the same number of protons but different numbers of neutrons in the nucleus.

Carbon-11, carbon-12 and carbon-14 are three isotopes of carbon. Of these, only carbon-12 is stable. It has equal numbers of protons and neutrons. The graph shows the relationship between the number of neutrons (N) and the number of protons (Z) for stable nuclei.

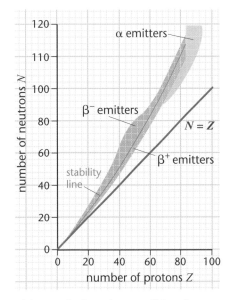

It can be seen from this graph that the condition for a nucleus to be stable depends on the number of protons:

- for values of Z up to 20, a stable nucleus has equal numbers of protons and neutrons
- for values of Z greater than 20, a stable nucleus has more neutrons than protons.

For stable nuclei with more than 20 protons, the neutron:proton ratio increases steadily to a value of around 1.5 for the most massive nuclei.

Unstable nuclei above the stability line in the diagram are **neutron-rich**; they can become more stable by decreasing the number of neutrons. They decay by β^-

emission; this leads to one less neutron and one extra proton and brings the neutron–proton ratio closer to, or equal to, one. An example is:

$$^{24}_{11}Na \rightarrow ^{24}_{12}Mg + ^{0}_{-1}e$$

Unstable nuclei below the stability line decay by β^+ emission; this increases the neutron number by one at the expense of the proton number. An example is:

$$^{11}_{6}C \rightarrow ^{11}_{5}B + ^{0}_{+1}e$$

The emission of an alpha particle has little effect on the neutron–proton ratio for isotopes that are close to the $N=Z$ line and is confined to the more massive nuclei. For these nuclei, emission of an alpha particle changes the balance of the proton–neutron ratio in the favour of the neutrons. In the decay of thorium-228 shown below, the neutron–proton ratio increases from 1.53 to 1.55.

$$^{228}_{90}Th \rightarrow ^{224}_{88}Ra + ^{4}_{2}He$$

The $N=Z$ line corresponds to a neutron:proton ratio of 1. As an alpha particle consists of two neutrons and two protons, its emission would hardly affect a neutron:proton ratio that is nearly 1.

Progress check

1 What is the effect on atomic and mass number of decay by:
  a an alpha particle only
  b a beta-minus particle only
  c a gamma photon only?

2 Technetium-99 is a gamma emitter with a half-life of 6 hours.
  A fresh sample is prepared with an activity A.
  a Calculate the activity of the sample when 24 hours have elapsed since its preparation.
  b Calculate the decay constant of technetium-99.

3 Complete the equation for the decay of phosphorus–32 by beta-minus emission.
  $^{32}_{15}P \rightarrow S+$

3 $^{32}_{15}P \rightarrow ^{32}_{16}S + ^{0}_{-1}e$

2 a $A/16$
  b $3.2 \times 10^{-5}\ s^{-1}$

1 a The atomic number decreases by 2 and the mass number decreases by 4.
  b The atomic number increases by 1 and the mass number is unchanged.
  c Both the atomic number and the mass number are unchanged.

4.3 Energy from the nucleus

LEARNING SUMMARY

After studying this section you should be able to:

- represent nuclear reactions by nuclear equations
- explain the equivalence of mass and energy
- describe how energy from the nucleus is released in a fission reactor

The nucleus

AQA A	AS	OCR A	M4
AQA B	AS	NICCEA	M6
EDEXCEL A	AS	WJEC	AS
EDEXCEL B	M5		

You will not find El in the periodic table – it is fictitious.

If the size of an atom is compared to that of a cathedral or a football stadium, the nucleus is about the size of a tennis ball. Evidence of the size of the nucleus comes from alpha particle scattering experiments (see section 4.4).

There are two types of particle in the nucleus, protons and neutrons.

> **KEY POINT**
>
> The nucleus of an element is represented as ^A_ZEl.
>
> Z is the atomic number, the number of protons.
>
> A is the mass number, the number of nucleons (protons and neutrons).

The element is fixed by Z, the number of protons. A neutral atom has equal numbers of protons in the nucleus and electrons in orbit. Different atoms of the same element can have different values of A due to having more or fewer neutrons. As this does not affect the number of electrons in a neutral atom, the chemical properties of these atoms are the same. They are called isotopes of the element.

The most common form of carbon, for example, is carbon-12, $^{12}_6$C, which has six protons and six neutrons in the nucleus. Carbon-14, $^{14}_6$C, has the same number of protons but two extra neutrons. These two forms of carbon are isotopes of the same element.

Atomic mass and energy conservation

AQA A	M4	OCR A	M4
AQA B	M5	NICCEA	M5
EDEXCEL B	M5	WJEC	M5

The phrase 'relative to the value' means compared to the actual amount of charge, ignoring the sign.

The charges and relative masses of atomic particles are shown in the table. The masses are in atomic mass units (u), where $1u = 1/12$ the mass of a carbon-12 atom $= 1.661 \times 10^{-27}$ kg. The charges are relative to the value of the electronic charge, $e = 1.602 \times 10^{-19}$ C.

atomic particle	mass	charge
proton	1.0073	+1
neutron	1.0087	0
electron	5.49×10^{-4}	−1

A carbon-12 nucleus consists of six protons and six neutrons. The mass of the atom is precisely $12\,u$, by definition, so after taking into account the mass of the electrons, that of the nucleus is $11.9967\,u$. The mass of the constituent neutrons and protons is:

$$6\,m_p + 6\,m_n = 6(1.0073\,u + 1.0087\,u) = 12.0960\,u$$

The nucleus has less mass than the particles that make it up. This appears to contravene the principle of conservation of mass. Einstein established that *energy has mass*. The mass that you gain due to your increased energy when you walk upstairs is infinitesimally small, but at a nuclear level this mass cannot be ignored. Any change in energy is accompanied by a change in mass, and *vice versa*.

Try using Einstein's equation to calculate the increase in your mass when you climb upstairs to see how small it is.

KEY POINT

Einstein's equation relates the change in energy, ΔE, to the change in mass, Δm

$$\Delta E = \Delta mc^2$$

where c is the speed of light.

It applies to **all** energy changes.

When dealing with the nucleus and nuclear particles, energy and mass are so closely linked that their equivalence, and that of their units, has been established.

KEY POINT

1 u = 931.5 MeV

where 1 eV (one electron volt) is the energy transfer when an electron moves through a potential difference of 1 volt.

Using this relationship, the separate conservation rules regarding mass and energy can be combined into one so that (mass + energy) is always conserved in nuclear interactions.

To split a nucleus up into its constituent nucleons requires energy. It follows that a nucleus has less energy than the sum of the energies of the corresponding number of free neutrons and protons. So the fact that a nucleus has less energy than its nucleons would have in isolation means that it also has less mass.

A common misconception is that the binding energy is the energy that holds a nucleus together. It is the energy needed to split it up.

KEY POINT

The difference between the sum of the masses of the individual nucleons and the mass of the nucleus is called the **mass defect** or **nuclear binding energy**. It represents the energy required to separate a nucleus into its individual nucleons.

In the case of carbon-12 the mass defect, or nuclear binding energy, is equal to $0.093\,u = 89.3$ MeV.

As would be expected, the greater the number of nucleons, the greater the binding energy. The diagram shows how the binding energy per nucleon varies with nucleon number. The most stable nuclei have the greatest binding energy per nucleon.

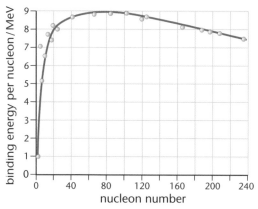

Joining and splitting

The nuclear reaction that releases energy in stars like our Sun is fusion.

In a fusion reaction:

• small nuclei join together to form larger, more massive ones
• the nucleus formed in fusion has less mass than those that fuse together, due to its greater binding energy per nucleon
• the mass difference is released as energy.

Fusion reactions can only take place at high temperatures where the nuclei have enough thermal energy to come together despite the repulsive forces that act as they approach each other.

Large nuclei release energy and become more stable by fission.

In fission:

- a large nucleus splits into two smaller ones and two or three neutrons
- the particles formed from fission have less mass than the original nucleus, due to the greater binding energy per nucleon
- the mass difference is released as kinetic energy of the fission fragments.

Induced fission

In a neutron-rich nucleus the neutron–proton ratio is greater than that required for stability.

A thermal neutron has energy similar to the mean kinetic energy of neutrons at room temperature.

Fission can occur spontaneously in a neutron-rich nucleus or it can be caused to take place by changing the make-up of a nucleus so that it is neutron-rich. This is known as induced fission.

Induced fission is triggered when a large nucleus absorbs a neutron. The nucleus formed has a greater mass than the original nucleus and neutron put together, so the neutron must supply the difference in the form of kinetic energy. A slow-moving, or thermal, neutron has the right amount of energy to cause the fission of uranium-235.

When uranium-235 undergoes fission:

- a nucleus of $^{235}_{92}U$ absorbs a neutron to become $^{236}_{92}U$
- the resulting nucleus is unstable and splits into two approximately equal-sized smaller nuclei, together with two or three neutrons
- a large amount of energy, approximately 200 MeV, is released as kinetic energy.

The fission of uranium-235 is illustrated in the diagram.

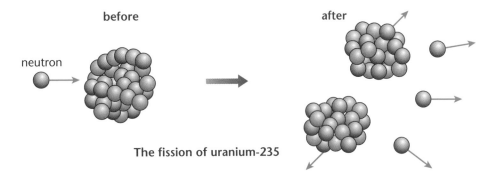

The fission of uranium-235

Check that this equation is balanced in terms of both mass and charge.

When uranium-235 fissions, the nuclei formed are not always the same. The equation represents one possible reaction, the products being barium and krypton.

$$^{235}_{92}U + ^{1}_{0}n \rightarrow ^{144}_{56}Ba + ^{90}_{36}Kr + 2^{1}_{0}n$$

Releasing the energy

If the neutrons released from fission go on to cause further fissions, then a chain reaction can build up. This is illustrated in the diagram overleaf.

If every neutron produced were to cause the fission of another uranium-235 nucleus, the reaction would quickly go out of control. This does not happen with a small quantity of uranium, since the neutrons produced in the fission process are moving at high speeds. These high-speed neutrons:

- are not as likely to be absorbed by uranium-235 nuclei as thermal neutrons are

- have a good chance of leaving the material altogether due to the relatively large surface of a small mass of material.

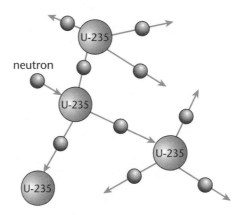

A chain reaction

The atomic bomb dropped on Hiroshima was detonated by combining three small masses of uranium into one that was above the critical mass.

For a chain reaction to be sustained then, on average, the neutrons released from each fission must go on to cause one further fission. The minimum mass of material needed for this to happen is called the **critical mass**. For uranium-235, the critical mass is 15 kg. This amount of uranium-235 in a spherical shape (so that it has the minimum surface area) can just sustain a reaction without it dying out.

Harnessing the energy

| AQA A | M4 | NICCEA | M5 |
| AQA B | M5 | WJEC | M5 |

In a **nuclear reactor**, energy released from fission is removed and used to generate electricity. Three important features of a reactor are the **moderator**, **control rods**, and **coolant**. These are shown in the diagram.

This type of reactor is called a thermal reactor as it uses thermal neutrons to cause fission.

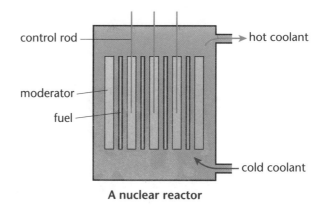

A nuclear reactor

The moderator:

- slows down the high-speed neutrons released by fission so that they are more likely to cause further fissions
- should not absorb neutrons.

Deuterium, ^2_1H, is an isotope of hydrogen that contains an extra neutron.

Graphite and heavy water, similar to normal water but using deuterium in place of hydrogen, are commonly used as moderators. Normal water is unsuitable as the hydrogen nuclei absorb neutrons.

The control rods:

- are used to control the neutron concentration by absorbing neutrons
- can be raised or lowered within the moderator material to increase or decrease the rate at which fissions take place
- must be able to absorb neutrons without becoming unstable.

The material used for control rods needs to have nuclei that are successful at capturing neutrons. Cadmium and boron are both used as they can capture neutrons over a large cross-section of their nuclei.

The coolant:

- should flow easily
- should not corrode the metal casings of the moderator and fuel rods
- removes energy to a heat exchanger, where steam is generated to turn turbines as in a coal-fired power station.

The coolant used depends on the temperature of the reactor core. Coolants in use include water, carbon dioxide and liquid sodium.

Nuclear fuels

Fast-breeder reactors use **fast** neutrons and **breed** their own fuel.

The first nuclear power stations to be built used natural uranium as a fuel. Natural uranium contains less than 1% uranium-235, the remainder being uranium-238 which does not fission readily. Enriched uranium fuel contains a higher proportion of uranium-235, up to 3%, so it has a longer usable lifetime than natural uranium. Fast-breeder reactors use the plutonium, $^{239}_{94}Pu$, formed when uranium-238, $^{238}_{92}U$, captures a neutron and then decays by β^- emission (twice). Plutonium needs fast neutrons to fission, but the spare neutrons left over from the fission process are used to create more plutonium fuel from uranium-238, so they make their own fuel from the relatively abundant isotope of uranium, uranium-238.

Safety

No nuclear reactor is totally safe. The coolant and other materials used in building the reactor become radioactive. The by-products of fission are also radioactive. Removal of the control rods results in a melt-down, as happened at Chernobyl. This released vast quantities of radioactive material into the atmosphere, having disastrous effects in the immediate vicinity and affecting farmland thousands of miles away.

In normal use a thick concrete wall built around the reactor shields the outside environment by absorbing much of the radiation emitted within. In an emergency, a reactor can be shut down by lowering the control rods fully so that they absorb more neutrons, leaving insufficient neutrons to maintain the chain reaction.

The way in which active waste materials are disposed of depends on the level of radioactivity.

- Low-level waste, such as laboratory clothing and packaging, is buried underground or at sea.

Spent fuel rods contain highly radioactive materials with long half-lives. There is no safe way of disposing of these.

- Medium-level waste, such as empty fuel casing and parts of the reactor fabrication, is kept in a concrete-lined store or underground cavern.
- High-level waste, such as the radioactive fission by-products and spent fuel rods, is kept in steel tanks of water to keep it cool.

Artificial nuclides

Many naturally-occurring radioactive isotopes have long half-lives and so are not suitable for use as tracers in the environment or in the body. Man-made radioactive isotopes can be produced in two ways:

- in a nuclear reactor, by irradiating a stable isotope with neutrons
- by firing charged particles at a stable isotope using a particle accelerator.

The resulting change in the nucleus is known as artificial transmutation. An isotope used extensively in medicine is a form of technetium-99, $^{99}Tc^m$, which emits gamma radiation only and has a half-life of 6 hours. It is formed when molybdenum-99, artificially created in a reactor, decays by beta-emission. This form of technetium-99 is particularly useful in medical diagnosis because:

- the gamma radiation can be detected outside the body using a gamma camera
- it does not emit the more intensely-ionising alpha and beta radiations which could cause cell damage
- the half-life is long enough for it to be detected for several hours after being injected and allowed to circulate the body
- the half-life is short enough for a minimum dose to be used
- by attaching the technetium-99 to different substances, it can be targeted towards specific body organs.

> The shorter the half-life, the greater the rate of decay so the smaller the dose needed to be detectable.

Progress check

1 4.0×10^5 J of energy is transferred to some water in a kettle to bring it to the boil.
Calculate the increase in mass of the water due to this energy transfer.

2 Describe the difference between nuclear *fusion* and nuclear *fission*.

3 Explain why a fast-breeder reactor does not need a moderator.

3 Fission is caused by fast neutrons, so they do not need to be slowed down.
2 Fusion is the joining together of two nuclei.
Fission is the splitting up of a nucleus.
1 4.4×10^{-12} kg

4.4 Probing matter

After studying this section you should be able to:

- *describe how the scattering of alpha particles and high-energy electrons gives evidence for the atomic model*
- *understand how electron diffraction enables measurements of atomic spacing to be made*
- *understand how the diffraction of X-rays and neutrons gives evidence for crystal structures*

LEARNING SUMMARY

Evidence for the atomic model

AQA A	M5	OCR B	M5
EDEXCEL A	AS	NICCEA	M5
EDEXCEL B	M4	WJEC	M5
OCR A	M4		

An alpha particle is a positively-charged particle consisting of two protons and two neutrons.

The experiments were carried out in a vacuum so that the alpha particles were not scattered by air particles.

The simple atomic model pictures the atom as a tiny, positively-charged nucleus surrounded by negatively-charged particles (electrons) in orbit. Evidence for this model comes from the scattering of alpha particles as they pass through a thin material such as gold foil. The results of these experiments, first carried out under the guidance of Rutherford in 1911, can be summarised as:

- most of the alpha particles travel straight through the foil with little or no deflection
- a small number are deflected by a large amount
- a tiny number are scattered backwards.

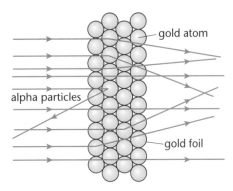

Rutherford concluded that the atoms of gold are mainly empty space, with tiny regions of concentrated charge. This charge must be the same sign as that of alpha particles (positive) to explain the back-scattering as being due to the repulsion between similar-charged objects.

The size of the nucleus

AQA A	M5	OCR A	M4
EDEXCEL A	AS	NICCEA	M6

It is a maximum value as the nucleus cannot be larger than this but is probably smaller, as the repulsive force reverses the direction of motion of the alpha particles before they reach the nucleus.

Rutherford's experiments established the existence of a nucleus that is very small compared to an atom. They also enabled the first estimates of the size of a nucleus to be made. When an alpha particle is scattered back along its original path:

- as it approaches the nucleus its kinetic energy is transferred to potential energy due to its position in the electric field
- at the closest distance of approach all its energy is potential
- the energy transfer is reversed as it moves away from the nucleus.

The closest distance of approach to the nucleus is therefore calculated from the kinetic energy of the alpha particle, assuming that Coulomb's law (see section 3.2) applies to the nucleus. This gives a maximum value for the size of the gold nucleus of about 10^{-14} m.

Electron diffraction

AQA A	M5	OCR B	AS
EDEXCEL A	AS	NICCEA	M5
OCR A	M4		

More precise evidence comes from the scattering of electrons, known as electron diffraction. Moving electrons show wave-like behaviour (see section 2.5) with a wavelength that depends on their speed.

Electrons accelerated through a potential difference of a few hundred volts have a wavelength similar to that of X-rays and gamma rays.

This wavelength is also similar to the spacing of the atoms in crystalline materials, so these materials provide suitable sized 'gaps' to cause diffraction.

Diffraction patterns formed by a beam of electrons after passing through thin foil or graphite show a set of 'bright' and 'dark' rings on photographic film, similar to those formed by X-ray diffraction (see below).

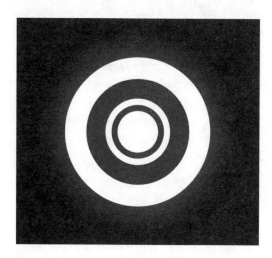

Analysis of these patterns is used to calculate the spacing between rows of atoms in the sample. This is typically around 10^{-10} m.

Higher energy electrons are used to estimate the size of a nucleus. Electrons accelerated through a voltage of a few hundred million volts have a wavelength comparable to that of a nuclear diameter.

> When probing matter, the choice of particle used depends on its interaction with the matter being probed.

Unlike alpha particles, electrons are not repelled as they approach a nucleus. Instead, they are diffracted in the same way that light is diffracted around a circular or spherical obstacle and a diffraction pattern similar to that shown in the diagram above emerges, with the rings having fuzzy edges.

Analysis of the diffraction pattern gives a nuclear size of about 10^{-15} m and the fuzziness of the rings shows that the nucleus does not have a perfectly circular cross-section.

The results of electron diffraction experiments show that:

> Unlike their nuclei, the atoms of different elements can have very different densities.

- all nuclei have approximately the same density, about 1×10^{17} kg m^{-3}
- the greater the number of nucleons, the larger the radius of the nucleus
- the nucleon number, A, is proportional to the cube of the nuclear radius, R.

An alternative way of writing this last point is:

the nuclear radius, R, is proportional to the cube root of the nucleon number, A.

> **KEY POINT**
>
> The relationship between nucleon number, A, and nuclear radius, R, is:
> $$R = r_0 A^{1/3}$$
> where r_0 has the value 1.2×10^{-15} m.

This relationship means that if the nuclei of ^{64}Cu and ^{32}S are compared, the radius of the copper nucleus is $2^{1/3} = 1.26$ times that of the sulphur, as the nucleon number of copper is twice that of sulphur. Similarly, the copper nucleus has a radius which is $4^{1/3} = 1.59$ times that of ^{16}O.

X-ray diffraction

OCR A   M4   WJEC   M5
NICCEA   M5

Metals are made up of small crystals. You can see these clearly on a clean metal surface such as the zinc plating on the inside of a can used for pineapple.

The diffraction of X-rays by crystalline materials such as metals shows the arrangement of atoms and allows measurements of their size. The regular arrangement of atoms in parallel planes causes a diffraction pattern when the waves used are comparable to the spacing. X-rays with a wavelength of the order of 1×10^{-10} m (0.1 nm) are suitable for this use. The diagram shows X-rays reflected at three consecutive planes within a crystal.

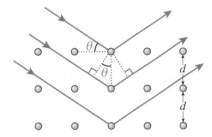

Constructive interference between the reflected X-rays is detected on photographic film. The angles, θ, at which this occurs are related to the wavelength of the X-rays and the spacing of the atomic planes, d, allowing the spacing to be calculated.

X-rays are absorbed by some metals due to their interaction with electrons.

X-rays do not interact with the nuclei of atoms in a material as they are uncharged and are not affected by electric or nuclear forces.

Neutron diffraction

OCR A   M4

Neutrons moving at a speed of around 4×10^4 m s^{-1} have a similar wavelength to that of X-rays and therefore should show a similar diffraction pattern of a crystal. This turns out to be the case, but neutron diffraction can give different information about a crystal structure to that obtained from X-ray diffraction. This is because:

The magnetic field is due to the spin of the neutrons and the charges contained within them.

* moving neutrons have a magnetic field
* this magnetic field interacts with the magnetic fields of orbiting electrons, revealing information about the magnetic structure of a material
* some metals absorb X-rays but transmit neutrons.

The diagrams on the next page compare X-ray and neutron diffraction patterns for the same metal alloy. Note that each one shows the molecules arranged in a regular pattern and reveals the structure of the crystal.

These diffraction patterns are due to a metal oxide, $ZrTiO_4$.

The contrasts in the diffraction patterns are:

* both show the arrangement of four oxygen atoms around the metal atoms
* the X-ray diffraction pattern shows the metals zirconium (Zr) and titanium (Ti) very clearly
* the neutron diffraction pattern shows the oxygen atoms more clearly than the metal atoms.

Neutron diffraction is used in preference to X-ray diffraction when examining the structure of nuclear fuel rods and some archaeological artefacts.

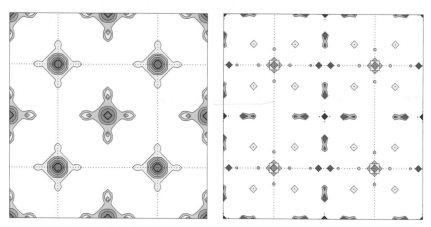

X-ray (left) and neutron (right) diffraction patterns for a metal alloy

Progress check

1 In alpha particle scattering by a gold nucleus, what feature of the results of the experiment shows that:
   a most of the atom is empty space
   b the nucleus has the same charge as the alpha particles?

2 What is the speed of an alpha particle at its closest distance of approach to a nucleus?

3 Explain why it is more difficult to accelerate neutrons than it is to accelerate electrons.

3 Electrons are charged so they can be accelerated by an electric field. Neutrons are uncharged.

2 Zero.

b Some alpha particles are back-scattered.

1 **a** Most of the particles pass through undeviated.

Radon-220 decays by *alpha-emission* to polonium-216.

(a) Explain what is meant by *alpha-emission*. [2]

> The nucleus emits a particle (1 mark) which consists of two protons and two neutrons (1 mark).

Note that the two marks awarded here are for describing the structure of an alpha particle and describing the emission.

(b) Complete the decay equation. [2]

$$^{220}_{86}Rn \rightarrow Po + He$$

> $^{220}_{86}Rn \rightarrow ^{216}_{84}Po + ^{4}_{2}He$ (1 mark for each correct symbol)

The nuclear equation should be balanced in terms of both mass (top number) and charge (lower number).

(c) The graph shows how the activity of a sample of radon-220 changes with time.

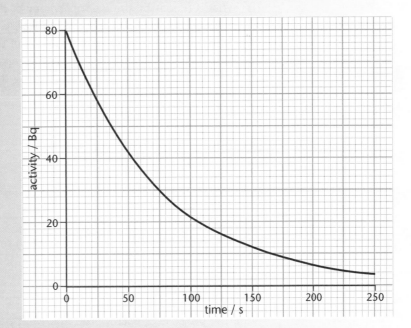

(i) Use the graph to determine the half-life of radon-220. [2]

> time taken for activity to halve from 80 Bq to 40 Bq = 52.5 s   1 mark
> time taken for activity to halve from 40 Bq to 20 Bq = 52.5 s
> half-life = average of these readings = 52.5 s   1 mark

It is important to take at least two readings here to determine the *average* time for the rate of decay to halve. Had the two readings been different, a third one would be necessary.

(ii) The relationship between half-life, $t_{1/2}$ and decay constant, λ, is:

$$\lambda \, t_{1/2} = \ln 2$$

Calculate the decay constant of radon-220. [2]

> $\lambda = \ln 2 \div t_{1/2}$ (1 mark) $= \ln 2 \div 52.5$ s $= 1.32 \times 10^{-2}$ s^{-1} 1 mark

If these relationships are not given in the question, they will be provided on a data sheet. Always check a data sheet, it may contain the formula that you cannot remember!

(iii) The activity, A, of a radioactive sample is related to the number of undecayed nuclei, N, by the formula:

$$A = \lambda N$$

Calculate the number of undecayed nuclei at time = 0 on the graph. [2]

> activity at time t = 0 is 80 Bq   1 mark
> $N = A \div \lambda = 80$ Bq $\div 1.32 \times 10^{-2}$ s^{-1} = 6060   1 mark

Practice examination questions

1

A sample of an ideal gas has a volume of 2.20×10^{-6} m^3 at a temperature of 25°C and a pressure of 2.02×10^5 Pa.

(a) Calculate the temperature of the gas in kelvin. [1]

(b) Calculate the number of moles of gas in the sample. [3]

$R = 8.3$ J mol^{-1} K^{-1}

(c) The absolute temperature of the gas is doubled, while the pressure remains constant.

(i) What is meant by the *internal energy* of the gas? [1]

(ii) In what form is this energy in an ideal gas? [1]

(iii) What happens to the internal energy of the gas when its absolute temperature is doubled? Explain your answer. [2]

(iv) Calculate the new volume occupied by the gas after its absolute temperature has doubled. [2]

2

In a nuclear reactor uranium-238, $^{238}_{92}$U, undergoes fission when it captures a neutron.

(a) (i) Write down the symbol for the isotope formed when uranium-238 captures a neutron. [1]

(ii) Explain what happens when a nucleus of this isotope undergoes fission. [3]

(iii) Explain how fission can lead to a chain reaction. [2]

(b) (i) Explain why nuclear fission results in the release of energy. [2]

(ii) In what form is the energy released? [2]

(iii) Outline how the energy released in fission is used to generate electricity. [3]

(c) State the purpose of each of the following parts of a nuclear reactor:

(i) control rods [1]

(ii) moderator. [1]

3

A kettle is filled with 1.50 kg of water at 10°C.
The kettle is fitted with a 2.40 kW heater.
The specific heat capacity of water is 4.2×10^3 J kg^{-1} K^{-1}.

(a) Calculate the time it takes for the water to boil after the kettle is switched on. [3]

(b) Suggest three reasons why the time taken is longer than the answer to (a). [3]

4

Iodine-123 and iodine-131 are two radioactive isotopes of iodine.
Iodine-123 emits gamma radiation only with a half-life of 13 hours.
Iodine-131 emits beta and gamma radiation and has a half-life of 8 days.

Both can be used to monitor the activity of the thyroid gland by measuring the count rate using a detector outside the body.

(a) Which of the radioactive emissions from iodine-131 can be detected outside the body? Give the reason for your answer. [2]

(b) In terms of the radiation emitted, explain why iodine-123 is often preferred to iodine-131. [2]

(c) Explain why a smaller dose is needed when iodine-123 is used than for iodine-131. [2]

(d) A patient is given a dose of iodine-123. After what time interval can medical staff be sure that the activity of the patient's thyroid gland is less than 2% of its greatest value? [2]

5

Information about the size of a nucleus can be obtained by firing high-energy electrons at the nucleus and examining the subsequent diffraction pattern.

(a) (i) Why are electrons suitable for this purpose? [2]
    (ii) Explain why the electrons need to be high-energy. [3]

(b) The relationship between the radius, R, of a nucleus and its atomic number, A is: $$R = r_0 A^{1/3}$$

where r_0 has the value 1.20×10^{-15} m.

Calculate the radius of the nuclei of:

(i) $^4_2$He [2]

(ii) $^{32}_{16}$S [1]

(c) Explain what the answers to (b) show about the densities of the helium and sulphur nuclei. [3]

6

The diagram shows some of the results of firing alpha particles at gold foil.

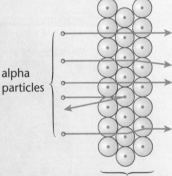

alpha
particles

atoms in gold foil

Explain how each of the following provides evidence for the atomic model.

(a) Most alpha particles pass through the foil undeviated. [2]

(b) A very small number of alpha particles are scattered backwards. [3]

Chapter 5

Medical physics

The following topics are covered in this chapter:

- *Body mechanics*
- *Using lenses*

- *Hearing*
- *Medical imaging and treatment*

5.1 Body mechanics

After studying this section you should be able to:

- *understand how bones and muscles enable body movement*
- *apply the principle of moments to bones and muscles*
- *explain how the force between the body and the ground changes when the body moves*

Body structure

OCR A ▸ MS.2

Animal cells do not have cell walls to give them rigidity, instead they rely on a system of **bones** which are connected together to make a **skeleton**. In the skeleton:

- bones meet at different types of **joint** which allow restricted movement
- at a joint, **ligaments** connect the bones together
- **muscles** are responsible for moving bones, they are attached to the bones by **tendons**.

The diagram below shows the ligaments, muscles and tendons at a hip joint.

> Muscles can only pull. They exert a force by contracting and release it when they relax. The two sets of muscle tissue shown in the diagram allow movement both ways. They are known as antagonistic muscles.

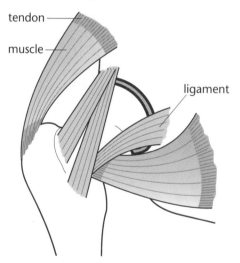

Bones are moved when muscles contract. The movement of a bone by a muscle is an example of a **lever** being used to magnify distance. A small contraction of the muscle causes a much larger movement at the far end of the bone that it connects to. The diagram below shows the turning forces involved when the biceps contract to raise the forearm from a horizontal position in order to raise a hand-held load.

> Use your own arm to estimate the distances between the forces and the pivot.

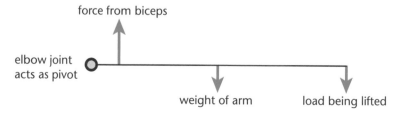

For the system to be balanced, the moment of the force from the biceps must be equal to the sum of the moments of the weight of the arm and the load held in the hand – this is an application of the **principle of moments**. Because of the relative distances, this means that the force from the biceps is much greater than the sum of the other two forces.

The **mechanical advantage** of a lever is a measure of how much easier it enables a job to be done.

> Most pulley systems have mechanical advantages greater than 1, they enable a small input force to move a large load.

> mechanical advantage = load/effort
>
> where the effort is the input force and the load is the output force.
>
> Mechanical advantage is a ratio so it has no units.

Joints in the body have mechanical advantages less than one; their effect is to enable a small movement to cause a much larger movement.

Forces and the body

OCR A M5.2

When standing still, the upward force from the ground that acts on a body is equal to the weight of the body. Walking and running both require additional forces to accelerate and decelerate the body both vertically and horizontally.

When a foot leaves the ground:

> When you walk or run, you push down on the ground so that it pushes back up on you.

- an additional upward force acts to accelerate the body upwards in order to raise the centre of mass
- a forwards-directed force due to friction acts to accelerate the body forwards.

When a foot hits the ground:

- an additional upward force acts to decelerate the body vertically
- a backwards-directed force due to friction acts to decelerate the body in the forwards direction.

The resultant of the forces acting on a foot is called the **ground force**; it acts parallel to the leg. The diagram shows the ground force, G, when a foot hits and leaves the ground during walking.

> The reaction force, R, is equal to the body weight when standing still and greater than the body weight when walking and running.

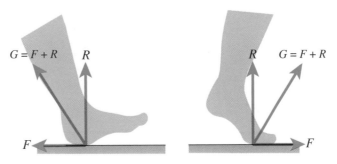

Bending and lifting

OCR A M5.2

Back injuries are a major cause of time lost at work due to illness.

When a person bends, the upper body turns around the base of the spine. The muscles attached to the spine have to produce a moment to counteract that of the weight of the upper body (about 60% of the total body weight). This is shown in the diagram below.

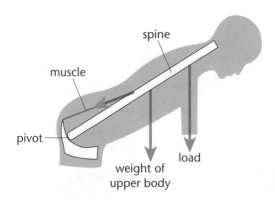

Because of the angles involved, the force from the muscle must be very much greater than the weight of the upper body. The diagram also shows what happens when a load is lifted. This puts a large stress on the muscle and the disc at the base of the spine. To avoid damage to these, heavy loads should be lifted by crouching, bending the knees and keeping the spine vertical.

Progress check

1 At a joint, what:
   a connects the bones together
   b attaches the muscles to the bones
   c contracts to cause movement?

2 The data in the table refers to the diagram of a forearm on page 120.

force	value of force	distance between force and pivot
biceps		0.04 m
weight of arm	20 N	0.18 m
load	50 N	0.36 m

Calculate the force in the biceps.

1 a ligament
   b tendon
   c muscle
2 540 N

122

5.2 Using lenses

After studying this section you should be able to:

- *use ray diagrams and the lens formula to describe image formation by lenses*
- *identify, and state the function of, the main parts of the eye*
- *explain how the quality of vision depends on the intensity of the ambient light*

The action of a lens

AQA A	M6B	OCR B	AS
OCR A	M5.2	NICCEA	AS

Light from a source converges after passing through a convex lens if the source is at a greater distance from the lens than the focal length.

The dotted lines in the diagram of a convex lens show where light **appears** to have come from.

The image is virtual in the case of a concave lens, and real for a convex lens.

When light leaves a lamp or other source of light it spreads out or diverges, reducing in intensity with increasing distance from the source. A concave or diverging lens exaggerates this effect, making the light more divergent. A convex or converging lens acts against this effect, so that the light either converges or becomes less divergent.

The diagrams show the effect of these lenses on a parallel beam of light.

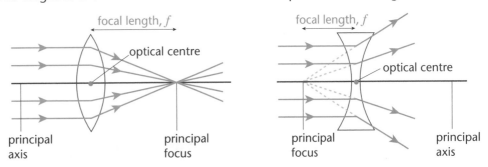

The diagrams also show some important features of the lenses:

- the **optical centre** is at the centre of the lens
- the **principal axis** is a line drawn through the optical centre, perpendicular to the plane of the lens
- the **principal focus** is the image position for light that is parallel to the principal axis
- the **focal length** is the distance between the optical centre and the principal focus.

The shorter the focal length of a lens, the greater the effect on converging or diverging light that passes through it. This effect is measured by the power of the lens.

> **Lens power** is calculated using the relationship:
>
> power = 1/focal length
>
> The unit of power is the dioptre (D) when the focal length is measured in m.

KEY POINT

The lens formula

AQA A	M6B	OCR B	AS
EDEXCEL B	AS	NICCEA	AS
OCR A	M5.2		

Concave lenses form **virtual images**. Convex lenses can form either **real** or virtual images, depending on the distance of the object from the optical centre of the lens compared to its focal length. In each case, the position, nature and size of the image can be determined by a ray diagram.

The diagram below shows ray diagrams for an object placed at a distance of 8 cm from converging and diverging lenses each of focal length, $f = 5$ cm.

A real image is formed if the light converges after passing through the lens. If the light is still diverging, then the image is virtual.

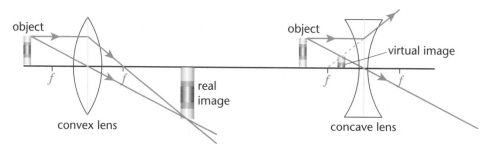

The position, nature and size of the images can be determined from the diagrams. They can also be calculated using the **lens formula**:

The magnification, i.e. the size of the image compared to the object, is equal to v/u.

> **KEY POINT**
>
> The lens formula applies to both convex and concave lenses. It relates the positions of the object and image to the focal length of the lens.
>
> $$\frac{1}{u} + \frac{1}{v} = \frac{1}{f}$$
>
> Where u is the distance between the object and the lens, v is the distance between the image and the lens, and f is the focal length of the lens.

This is known as the 'real is positive' sign convention.

When using the lens formula:

* real objects, images and the focal length of a convex lens have positive values
* virtual images, and the focal length of a concave lens, have negative values.

Structure of the eye

AQA A · M6B · NICCEA · AS
OCR B · AS

The eye uses two converging lenses, the cornea and the eye lens, to produce a real image on the retina. The diagram shows the main parts of the eye and the focusing action of these lenses.

The cornea also acts as a protective covering, so that no dust or other objects can enter the eye.

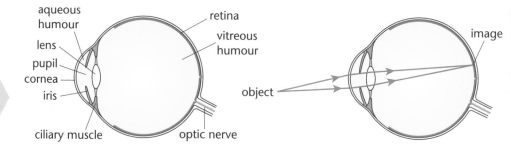

* The **cornea** is a fixed lens that carries out most of the convergence of light required for image formation.
* The **aqueous humour** and **vitreous humour** supply the eye with essential nutrients and exert pressure to maintain the shape of the eyeball.
* The **iris** adjusts the size of the pupil to control the amount of light entering the eye.
* The **lens** is of variable focal length and acts as a fine-focus control.
* The **ciliary muscles** are connected to the lens by ligaments: they can relax and contract to change the shape of the lens.
* The **retina** consists of nerve-endings which are sensitive to the colour and brightness of light and produce electrical signals.
* The **optic nerve** transfers these signals to the brain.

Focusing the eye

AQA A   M6B   NICCEA   AS
OCR B   M5.2

For a normal eye which does not require spectacles, the near point is at a distance of about 25 cm and the far point is at infinity.

The **near point** is the closest distance that an eye can focus on. The eye lens is at its most powerful when focusing on the near point and at its least powerful when focused on the **far point**, the greatest distance from the eye at which objects can be seen clearly. Whatever the distance the object is that the eye is focused on, objects nearer and further than this can also be seen clearly. The range of distances over which objects can be seen in focus is called the depth of field.

Adjustment of the eye to focus on objects at different distances is called **accommodation**; this is achieved by changing the shape of the lens.

Defective vision

AQA A   M6B   NICCEA   AS
OCR B   M5.2

An eye lens is elastic tissue which becomes more difficult to stretch with increased use.

The common defects in vision are:

- **short sight** or **myopia** when an eye cannot focus on distant objects but can focus on near objects. It is caused by the eyeball being too long or the lens, at its weakest, being too strong.
- **long sight** or **hypermetropia** is the opposite of short sight; distant objects can be seen clearly but near ones cannot. It is caused by the eyeball being too short or the lens, at its strongest, being too weak.
- **presbyopia** is the condition of being both long-sighted and short-sighted. It often occurs in older people due to the lens becoming stiffer with age.
- **astigmatism** is caused by the surface of the cornea not being spherical. This results in objects in one direction being sharp and in focus while those in a different direction are blurred.

The diagrams show the conditions of short sight and long sight and their correction with converging and diverging lenses respectively.

These diagrams shows light from a distant object arriving at a short-sighted eye.

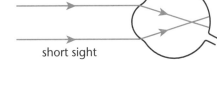

short sight

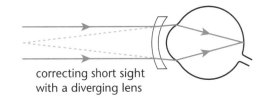

correcting short sight with a diverging lens

These diagrams show light from a near object arriving at a long-sighted eye.

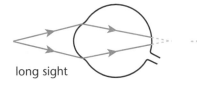

long sight

correcting long sight with a converging lens

The lens formula can be used to calculate the focal length, and hence power, of the correcting lens needed. The object distance, u, is the distance of the actual object and the image distance, v, is the desired distance for the eye to focus on it.

People who suffer from presbyopia often wear bifocal spectacles. The lower half of each lens is convex for reading and the upper half is concave for distant viewing.

To correct astigmatism cylindrical lenses are used. These are only curved in one direction to compensate for the unsymmetrical curvature of the cornea.

Colour vision

AQA A   M6B   NICCEA   AS
OCR B   M5.2

The light-sensitive cells that make up the retina have different sensitivities:

- **rods** are sensitive to light intensity only
- there are three types of **cones**, called red, green and blue. Each of these responds to a different range of wavelengths and frequencies.

This explains why, in low-intensity light such as moonlight, we see in monochrome.

Cones enable colour vision but they do not respond to low-intensity light. The graph shows the relative sensitivities of the three types of cone.

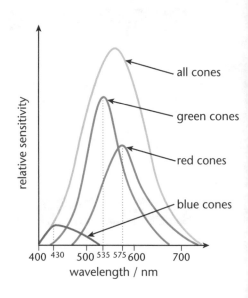

The low sensitivity of the blue cones explains why we never perceive this colour as bright as red or yellow. An advertising hoarding designed to catch the eye would not use blue to do so. However, it is a more acceptable colour for 'low-key' signs and buildings.

The diagram shows why blue-coloured lamps always appear dim and yellow-coloured lamps appear bright.

Vision in low-intensity light is scotopic – we cannot detect colour but only differences in brightness. In scotopic vision the cones are not responsive so the rods are the only detectors operating.

Vision is most acute at the fovea, or yellow spot. This is the point on the retina where the light-sensitive cells are most closely-packed.

In more intense light the cones take over and vision is photopic. Photopic vision allows differentiation of colour in addition to intensity. The ability of the eye to distinguish between two objects that are close together, called its spatial resolution, is also greater in photopic vision. Spatial resolution requires at least one unstimulated cell between those that are stimulated. Resolution is greatest at the centre of the retina where the cells are most concentrated. It is worst at the edge of the retina where the cells are more widespread and several rods share a single nerve fibre, so the brain cannot distinguish between individual rods.

Making still pictures move

AQA A   M6B

In film animation, a series of photographs is taken, with each one showing a small change in position from the previous photograph.

We 'see' moving pictures when watching a film at a cinema or watching television. In reality there are no moving pictures.

- At the cinema, 24 still pictures appear on the screen each second.
- On the television, there is only a 'splodge' of light, consisting of three coloured patches, at any instant. The splodge sweeps the screen 25 times each second, with each sweep consisting of 625 lines.

A succession of still pictures merge into each other because the light-sensitive cells on the retina remain in an excited state for a short time after the image has been removed. This is called persistence of vision and it also explains why we see a moving dot on a CRO screen as a line.

Progress check

1 An object of height 2.0 cm is placed on the principal axis at a distance of 12.0 cm from a convex lens of focal length 5.0 cm.
   By drawing or calculation, work out the position, size and nature of the image.

2 Which TWO parts of the eye converge light as it passes through them?

3 Explain why the brain does not perceive colour in low-intensity ambient light.

3 The cones do not function in low-intensity light, and the rods only detect brightness.
2 The cornea and the eye lens.
1 The image is real, 8.6 cm on the other side of the lens and 1.4 cm high.

5.3 Hearing

After studying this section you should be able to:

- describe how sound travels through the ear and is detected by the brain
- explain that the ear is sensitive to a range of frequencies and intensities of sound
- compare sound intensities in W m⁻² and in decibels

Ear structure

AQA A ▶ M6B   OCR A ▶ M5.2

The ear receives sound in the form of compression waves and produces an electrical signal that represents these waves. This is the action of a transducer, a device that transfers a signal from one form to another.

The diagram shows the main parts of the ear.

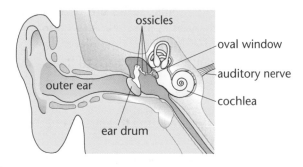

ossicles
oval window
auditory nerve
cochlea
outer ear
ear drum

The three ossicles are the malleus (hammer), incus (anvil) and stapes (stirrup).

The ear drum is also known as the tympanic membrane. It is very thin and vibrates with a typical amplitude of 10^{-9} m.

In the ear:

- sound waves in the outer ear cause the ear drum to vibrate
- the ossicles (ear bones) transmit the vibrations to the oval window of the cochlea. In doing so they reduce the amplitude of the vibrations and maximise the energy transfer
- vibrations of the fluid in the cochlea are detected by tiny hairs which produce electrical impulses
- these impulses are transmitted along the auditory nerve to the brain.

Sensitivity of the ear

AQA A ▶ M6B   NICCEA ▶ AS
OCR A ▶ M5.2

The ear responds to a range of frequencies and a range of intensities.

The intensity of a wave is defined as the power incident per unit area perpendicular to the direction of wave travel.

$$I = P/A$$

where intensity, I is measured in W m⁻². Intensity can also be measured in **decibels** (dB) – see below.

KEY POINT

Although increasing the intensity of a sound at a particular frequency makes it sound louder, we perceive sounds of different frequencies and the same intensity as varying in loudness. This is because of the different sensitivity of the ear at different frequencies, shown in the graph overleaf.

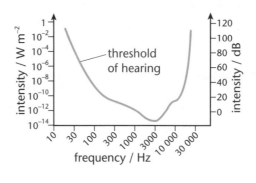

threshold
of hearing

> The threshold of hearing is the minimum intensity required to be able to detect a sound of that frequency.

This graph shows that:

- the frequency range of a human ear is approximately 20 Hz to 20 000 Hz, but the upper limit falls with increasing age and exposure to noise due to wear of the ossicles
- the minimum intensity required to detect a sound, I_0, depends on frequency
- the ear is at its most sensitive at frequencies around 3000 Hz.

> On a logarithmic scale, equal distances on the scale are used to represent multiplication by 10.

Logarithmic scales are used on the graph because the ear responds to relative changes in intensity, it has a logarithmic response. This means that doubling the intensity doubles the perceived loudness, no matter what the initial intensity of the sound was. This leads to the decibel scale for measuring intensity.

$$\text{intensity in decibels} = 10 \log (I/I_0)$$
where I_0, the threshold of hearing, is taken to be 1.0×10^{-12} W m^{-2}.

On the decibel scale, an increase of 10 dB corresponds to the intensity becoming ten times as great.

> Normal conversation has an intensity of about 1×10^{-6} W m^{-2} or 60 dB.

A sound of intensity 120 dB or 1 W m^{-2} causes discomfort as it causes feeling in the ear; one of 140 dB or 100 W m^{-2} causes extreme pain in the ear.

Loudness

AQA A · M6B · OCR A · M5.2

> By definition, loudness and intensity in decibels have the same numerical value at 1 kHz.

Because our ears are most sensitive to frequencies around 3000 Hz, higher and lower frequencies need a greater intensity to sound equally loud. Loudness is measured in phons, where 1 phon is equivalent to an intensity of 1dB at a frequency of 1kHz. The graph shows the variation in intensity required to cause equal loudness at four loudness levels.

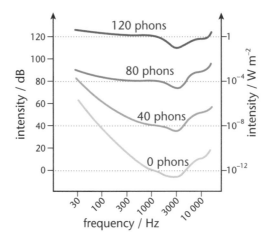

The dBA scale of intensity takes into account the fact that mid-range frequency sounds seem louder than those at the extreme frequencies.

Meters used to measure sound levels in dBA filter out the high and low frequencies so that more weighting is given to those that cause the loudest sounds.

Attenuation and impedance

The decrease in intensity of a wave as it travels through a material is known as attenuation. Attenuation is due to a number of factors, including:

- the energy of the wave becoming spread over a wider area as it travels from its source
- interactions with the material including absorption, diffraction and scattering.

Sound travels faster in solids than in liquids and faster in liquids than in gases.

The opposition of a material to sound passing through it is called acoustic impedance. The acoustic impedance depends on the density of the material and the speed at which sound travels in it. Maximising the transfer of sound from one material to another depends on matching the impedances as closely as possible.

Progress check

1 State two factors that reduce the upper limit of the hearing range.
2 A sound has an intensity of 5.0×10^{-7} W m^{-2}. Calculate the intensity in dB.

2 57 dB.
1 Age and exposure to noise.

5.4 Medical imaging and treatment

After studying this section you should be able to:

- *explain how the heart is controlled by electrical signals*
- *describe and compare the techniques used for imaging the body*
- *recall the methods used to calculate absorption of harmful radiation*

LEARNING SUMMARY

Biological measurement

AQA A M6B

The valves are necessary to ensure that blood does not flow from the ventricle to the atrium when the ventricle contracts.

The **heart** is a pump controlled by electrical signals. One side takes in de-oxygenated blood from the body and pumps it to the lungs; the other side takes in oxygenated blood from the lungs and pumps it to the body. The diagram shows the heart and the one-way valves that only allow blood to flow in the directions shown by the arrows, from the **atrium** to the **ventricle**.

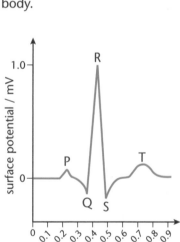

Electrical signals that control the action of the heart:

- are produced in the **sino-atrial node**
- are carried by nerve fibres
- cause contraction of the atria, followed by contraction of the ventricles to pump blood through the heart and round the body.

Body fluids also conduct the electrical signals, enabling them to be detected at the surface of the skin by electrodes. A display, such as that produced on a CRO of the change in voltage with time, is called an **electrocardiagram** (ECG).

In the ECG shown in the diagram:

- the pulse at P causes the atria to contract

The pulse that causes relaxation of the atria occurs at the same time as, and is masked by, the QRS pulse.

- the QRS pulses causes the ventricles to contract
- the T pulse causes the ventricles to relax.

Using ultrasound

AQA A M6B NICCEA AS
OCR A M5.2

The alternating voltages due to the reflected waves are interpreted by a computer to produce an image.

Ultrasound consists of compression waves with a frequency above that of human hearing. It is useful for examining the internal structure of the body because it passes through flesh, but some reflection takes place at tissue boundaries. These reflections are used to generate images.

Pulses of ultrasound are produced and detected by a **piezoelectric crystal**:

- an alternating voltage applied to the crystal causes it to vibrate
- the voltage is removed to allow the crystal to detect any reflections
- when the crystal is made to vibrate by the reflected pulses it generates an alternating voltage.

Reflections at tissue boundaries are due to the tissues having different acoustic impedances (see section 5.3). The greater the difference in acoustic impedance, the more intense the reflection. For this reason a jelly-like substance is spread over the patient's skin during scanning, to improve the impedance matching and maximise the amount of energy transmitted into the body.

Types of ultrasound images include:

- an **A-scan** (amplitude modulated) which uses the echoes to show how the tissue boundaries are related to depth below the surface of the skin, A-scans are used to make precise determinations of the position of an object such as a brain tumour

- a **B-scan** (brightness modulated) which gives a two-dimensional picture, such as that of a fetus.

Ultrasound scans have the advantage in that ultrasound is non-invasive and does not cause ionisation and damage to body tissue; their disadvantage is that the high frequencies needed for good resolution are attenuated more than lower frequencies, so they can only produce high-resolution scans close to the surface of the body.

> The scanning of a fetus is now a routine procedure carried out on all pregnant women to check for abnormalities and determine the precise date when birth is due to take place.

Producing X-rays

AQA A   M6 B   NICCEA   AS
OCR A   M5.2

X-rays are produced when high-speed electrons hit a target. In the X-ray tube shown in the diagram:

- electrons are emitted from the cathode by thermionic emission

- they are accelerated through the vacuum by the high-voltage anode

- when they strike the anode about 1% of the energy is transferred to X-rays, the rest heats the anode

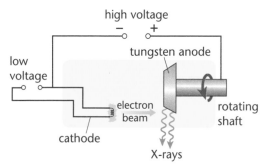

A rotating anode X-ray tube

- the anode rotates to avoid overheating of one small area.

> X-rays are emitted when electrons strike the screen of a television or cathode ray tube.

The **intensity** of the beam is determined by the filament current which causes thermionic emission. The energy of the X-ray photons is determined by the anode voltage; increasing this voltage produces higher-energy, shorter-wavelength X-rays which are more penetrative.

Using X-rays in diagnosis

AQA A   M6 B   NICCEA   AS
OCR A   M5.2

Whereas ultrasound images are produced from reflections, X-ray images rely on some absorption and some transmission by the body. As an X-ray beam passes through tissue it becomes attenuated (see section 6.3). The intensity varies exponentially with thickness.

> In exponential attenuation, the intensity of the beam is reduced by the same proportion in passing through equal thicknesses of material. This means that the half-thickness is independent of the initial intensity of the beam.

The intensity of an X-ray beam, I, after passing through a distance x of absorbing material, is given by:

$$I = I_0 e^{-\mu x}$$

Where I_0 is the initial intensity and μ is a constant, called the **linear attenuation coefficient**. The value of μ depends on the material and the energy of the X-rays.

The **half-value thickness** of a material, $x_{1/2}$, is the thickness that halves the original intensity of the beam. This relationship between μ and $x_{1/2}$, is:

$$\mu x_{1/2}, = \ln 2$$

The linear attenuation coefficient changes with changing density, ρ, of the material. The mass attenuation coefficient, μ_m, defined as $\frac{\mu}{\rho}$, is independent of density.

An X-ray photograph is a shadow picture. A detector placed behind the part of the body being X-rayed shows areas of varying greyness, with the dark areas representing the least attenuation of the beam. When bones such as teeth are X-rayed, they show up as white. Suitable detectors are:

- black and white photographic film, but this requires a beam of high intensity
- an **intensifying screen** which contains a material that absorbs energy from the X-rays and re-emits it as light, a process known as **fluorescence**. The light then produces a photographic image. Film is more sensitive to light than to X-rays, so less exposure is required.

> Studying a moving picture enables medical staff to diagnose faults in the way that organs act or bones move.

- a fluoroscopic image intensifier; the X-rays cause electrons to be ejected from a photocathode. These electrons are then accelerated in the same way as in a cathode ray tube, where they produce a bright image on a fluorescent screen. This technique is used when a 'moving' picture is required.

When intestines are X-rayed, there is very little absorption by the soft tissue and the result is a low-contrast image. The contrast between the intestines and surrounding tissue is increased by giving the patient a 'barium meal'. This lines the intestines with a material that is opaque to X-rays, so they show up as white.

A CT or CAT (**computerised axial topography**) scan uses a rotating X-ray beam to produce an image of a slice through the body. The beam is rotated around the patient as the patient moves through the beam. The scattered X-rays are detected and used by a computer to build up a picture. Three-dimensional images can be created from images of the individual slices. CAT scans detect slight changes in attenuation and are often used for detecting tumours and cancers.

Fibre optics

AQA A ⟩ M6B ⟩ OCR A ⟩ M5.2

> Fibres are normally clad with a glass of lower refractive index than the core in order to minimise the value of the critical angle.

An **optical fibre** can transmit light round bends and corners using **total internal reflection**. When light strikes the interface between the fibre core and its cladding it is reflected back inside the fibre provided that the angle of incidence is greater than the critical angle. This is shown in the diagram.

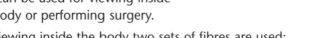

cladding
core

An **endoscope** is a bundle of fibres that can be used for viewing inside the body or performing surgery.

For viewing inside the body two sets of fibres are used:

- the fibres that take light into the body are **non-coherent**, they do not need to stay in the same position relative to other fibres
- the fibres that transmit the image out of the body are **coherent**, they are in the same positions relative to each other at each end of the bundle. The narrower the fibres used, the greater the resolution of the image.

> Tissue becomes heated by absorption of the energy from laser radiation.

Lasers are sources of high-intensity electromagnetic radiation that can be passed into the body along an optical fibre. This allows heating to be applied to small areas of tissue by absorption of the laser beam. Used in this way optical fibres can:

- act as a scalpel by cutting through tissue but without causing bleeding
- repair tears and stop bleeding by coagulating body fluids.

Magnetic resonance imaging

OCR A   M5.2   NICCEA   AS

Magnetic resonance imaging (MRI) detects the presence of hydrogen nuclei. Since body tissue has a high water content, it can be used to produce an image of the body. It works by applying pulses of a high-intensity and high-frequency magnetic field which results in the emission of electromagnetic radiation from the nuclei by the following process.

> Very powerful electromagnets are needed to produce the magnetic fields required by an MRI scanner.

- Hydrogen nuclei spin, giving them a magnetic field.
- When an external magnetic field is applied, they rotate around the direction of the field. This rotation is called **precession**, it is like that of a spinning top when it 'wobbles'.
- The frequency of precession is called the **Larmor frequency**.
- Application of a pulse of radiation at the Larmor frequency causes **resonance**, the nuclei absorb the energy into the precession.
- When the pulse is removed the nuclei lose the energy, emitting it as electromagnetic radiation. This takes place over a short period of time known as the **relaxation time**.

The emissions from the hydrogen nuclei are detected by a radio aerial and processed by a computer to give three-dimensional body imaging. MRI scanning is non-invasive and does not cause ionisation, but it is very expensive in terms of the capital equipment and running costs.

Using ionising radiation

OCR A   M5.2

> Hydrogen peroxide is extremely corrosive and can cause extensive damage when formed within a cell.

Ionising radiation such as X-rays and gamma rays:

- can damage enzymes and DNA by ionisation
- can cause the formation of hydrogen peroxide from water, which in turn causes cell damage.

The effects of this include skin burns, cancer and sterility. For this reason, **exposure** to ionising radiation should always be kept to a minimum.

> **KEY POINT**
>
> Exposure, X, is defined in terms of the ionisation that the radiation causes:
> $$X = \frac{Q}{m}$$
> where Q is the total charge of one sign of ion (either positive or negative) produced in a mass m of air. Exposure is measured in C kg^{-1}.

This definition of exposure only applies to X-rays and gamma radiation. The **absorbed dose** of radiation measures the energy absorbed by a material.

> **KEY POINT**
>
> Absorbed dose, D, is defined as
> $$D = \frac{E}{m}$$
> where E is the energy absorbed by mass m. Its unit is the gray (Gy) where 1 Gy = 1 J kg^{-1}.

The effect of absorbing radiation depends on the absorbed dose and the type of radiation involved. Alpha particles and neutrons are absorbed over a short distance whereas X-rays and gamma rays cause less damage for the same absorbed dose as their energy is distributed over a greater mass of material. The **quality factor** takes into account the effects of different radiations and is used to calculate equivalent effects of absorbing different radiations.

Effectively the dose equivalent is used to compare the energies absorbed by doses of different radiations.

The **dose equivalent**, H, of a radiation is defined by:
$$H = Q \times D$$
where Q is the quality factor of the radiation.
Dose equivalent is measured in sievart (Sv).

X-rays and gamma rays have a quality factor of 1, while that for radiation which is absorbed more readily is higher, that for alpha particles being 20.

The effects of a dose of ionising radiation can be classified as stochastic (random) or non-stochastic:

- stochastic effects such as cancer are due to cell mutations; no minimum dose is required but the greater the dose the more likely it is that the effect will occur
- non-stochastic effects such as skin burns need a minimum dose to occur; the greater the dose above this minimum value, the worse the effect.

When X-rays and gamma rays are used to kill a malignant tumour several beams are used, aimed at the patient in different directions. The diagram shows how the beams intersect at the tumour. This ensures that the tumour receives a high dose while that in the surrounding tissue is relatively low.

A similar effect is achieved by rotating a single beam around the patient so that the beam is always directed towards the tumour.

this area receives the greatest intensity of radiation

Progress check

1 In an X-ray tube:
   a what determines the intensity of the X-ray beam
   b what determines the wavelength of the X-rays produced?

2 Suggest TWO reasons why non-invasive examination is preferred to invasive examination.

3 Why is ultrasound preferred to X-rays for producing an image of a fetus?

3 Ultrasound does not cause ionisation or damage body tissue.

2 There are no wounds, so infections cannot occur.
The patient recovers more rapidly.

1 a The current in the filament.
   b The anode voltage.

Sample question and model answer

Ultrasound is used for examining delicate organs and tissue. The diagram shows how an ultrasound probe is used to examine the retina.

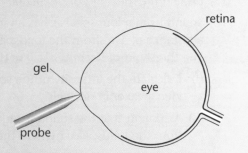

The depth of the eye is 2.5 cm.
The speed of sound in the eye tissue is 1.50×10^3 m s^{-1}.

(a) For the best results, the wavelength used for an ultrasound scan should be 1/200 of the depth of the tissue being scanned.
Calculate the optimum frequency for scanning a retina. [3]

This is an example of where a question on an extension topic requires recall of material from the core, in this case the wave equations.

optimum wavelength = 1/200 x 2.5 x 10^{-2} m = 1.25 x 10^{-4} m　　1 mark
optimum frequency, f = v/λ = 1.50 x 10^3 m s^{-1} ÷ 1.25 x 10^{-4} m　　1 mark
　　　　　= 1.20 x 10^7 Hz　　1 mark

(b) Explain why the gel is used between the probe and the eyeball. [2]

This is an example of impedance matching to maximise the energy transferred between two materials.

This maximises the energy transmitted into the eye (1 mark) by reducing the amount reflected (1 mark).

(c) In an ultrasound scan a piezoelectric crystal acts as both transmitter and detector of the ultrasound. It emits short pulses of waves, with a time gap between each pulse.

(i) Explain why there is a time gap between each pulse of waves. [2]

To allow the reflected waves to return to the crystal and be detected (1 mark) without interfering with the waves emitted by the crystal.　1 mark

(ii) Estimate the minimum time gap that should be left between pulses in this example. [3]

The minimum time should be that taken for the pulse to travel to the retina and for the reflection to be detected.　　1 mark

The factor of two is needed here because the pulse has to travel to the back of the eye and return.

time = distance ÷ speed = 2 x 2.50 x 10^{-2} m ÷ 1.50 x 10^3 m s^{-1}　1 mark
　　　　　= 3.33 x 10^{-5} s　　1 mark

Practice examination questions

1

Sound is attenuated as it travels in the air.

(a) Explain the meaning of *attenuation*. [1]

(b) Sound from a loudspeaker has an intensity of 6.0×10^{-5} W m^{-2}, measured at a distance of 1 m from the loudspeaker.
The threshold of hearing is 1.0×10^{-12} W m^{-2}.

(i) Calculate the intensity of the sound at a distance of 1 m from the loudspeaker in decibels. [3]

(ii) Assuming that the sound is radiated in all directions, calculate the distance from the loudspeaker where the sound becomes inaudible. [3]

(iii) What is the intensity, in dB, of a sound at the threshold of hearing? [1]

(c) The ear is most sensitive to sounds with a frequency around 3000 Hz.
Explain how the perceived loudness of sounds of equal intensity depends on their frequency. [3]

2

For ultra-sound scanning, a piezoelectric crystal produces a series of pulses of ultrasound and then detects the reflections.

An ultrasound scan is used to detect a tumour at a depth of 6.0 cm below the surface of the skin. The speed of sound in tissue is 1500 m s^{-1}.

(a) Explain whether an A-scan or B-scan is appropriate for this application. [2]

(b) Calculate the time that elapses between the pulse being sent and the echo being detected. [2]

(c) Explain why it is necessary to use a short pulse of ultrasound. [2]

(d) Explain why it is necessary to use pulses of ultrasound rather than a continuous wave. [2]

(e) The optimum frequency is one that produces a wave with a wavelength that is 1/200 the depth of the organ.
Calculate the optimum frequency for this application. [3]

(f) Before a scan is carried out, the skin of the patient is coated with a gel.
Explain the purpose of this gel. [2]

(g) CAT scans are also used to detect tumours.

(i) Outline the principle of operation of a CAT-scanner. [3]

(ii) Describe the advantages and disadvantages of using a CAT scan rather than an ultrasound scan to detect a tumour. [3]

Chapter 6

Astronomy and cosmology

The following topics are covered in this chapter:

- *Looking up*
- *The model unfolds*
- *The expanding Universe*
- *Classifying stars*
- *Relativity*

6.1 Looking up

LEARNING SUMMARY

After studying this section you should be able to:

- *describe the structure of reflecting, refracting and radio telescopes*
- *explain how the resolution of a telescope depends on its aperture*
- *recall that a charge-coupled device is used to give a brighter image*

Optical telescopes

AQA A   M6A

A **refracting telescope** uses two converging lenses. The objective lens produces a real image at its principle focus. The weaker this lens, the greater the size of the image produced and the longer the telescope. The eyepiece lens acts as a magnifying glass; by placing it so that its principal focus is at the position of the real image being magnified, the final image is formed 'at infinity' and the telescope is said to be in **normal adjustment**. This is shown in the diagram.

Refracting telescopes are limited in the diameter of the objective lens that can be used, since the weight of a large lens can cause the glass to flow and change its shape.

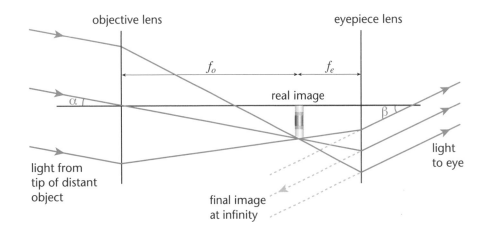

Since the size of the image on the retina depends on the angle subtended by the light reaching the eye, **angular magnification** is used to measure the factor by which the telescope increases this angle.

> **KEY POINT**
>
> Angular magnification, M, $= \dfrac{\text{angle subtended at eye by image}}{\text{angle subtended at eye by object}} = \dfrac{\beta}{\alpha}$
>
> For a telescope in normal adjustment, $M = \dfrac{f_o}{f_e}$

As with a converging lens, the principal focus is the point where light parallel to the principal axis converges.

In the Cassegrain **reflecting telescope** a convex mirror is used to converge parallel light from a distant object. This would produce a real image at its principal focus, but the light is intercepted by a small concave mirror which produces an enlarged image for magnification by the eyepiece lens. This is shown in the diagram.

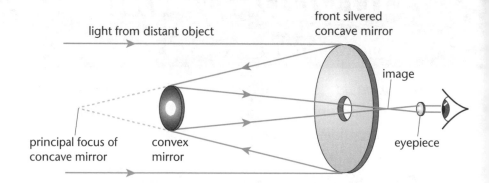

light from distant object

front silvered concave mirror

image

eyepiece

principal focus of concave mirror

convex mirror

A front-silvered mirror is used to prevent loss and possible distortion of light as it passes through the glass of a rear-silvered mirror.

Both types of telescope can produce blurring of the image called **aberration**.

- **Chromatic aberration** is due to light of different wavelengths being focused at different points; it is reduced by using combination lenses made with two different types of glass.

- **Spherical aberration** is caused by light that strikes the outer edges of a lens or mirror being brought to a focus slightly closer than those from the centre; it is reduced by using plano-convex lenses (only one face is curved) in refracting telescopes and parabolic mirrors in reflecting telescopes.

Resolution

AQA A   M6A

The **resolving power** of a telescope describes its ability to distinguish two stars that are close together, rather than identify them as a single star. When light passes through a circular aperture it is diffracted (see section 2.4). The bright central ring (known as the **Airy disc**) is surrounded by a number of low-intensity rings. The diagram shows the variation of intensity with angle from the centre of the Airy disc.

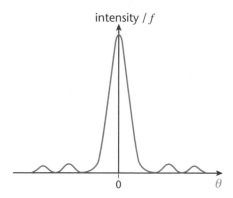

intensity / f

0

θ

This is the separation required for both Airey discs to be discernible.

To be able to distinguish between two stars:

- the minimum separation of the images is when the central maximum of each is at the same place as the first minimum of the other

The symbol ≈ means 'of the order of' rather than precisely equal to.

- the **Rayleigh criterion** for this to happen is that the angle between the stars $\theta \approx \dfrac{\lambda}{D}$,, where D is the diameter of the telescope and θ is in radians

- θ is known as the **resolving power** of the telescope.

Improving the image

AQA A   M6A

Charge-coupled devices are cheap to manufacture but they are limited in size.

Images that appear dim when viewed through a telescope can be viewed as much brighter images by storing them on a **charge-coupled device**:

- A charge-coupled device is a thin slice of silicon whose upper surface is divided into a large number of pixels

- When a photon of light is absorbed by a pixel, the energy releases an electron from the silicon onto the pixel.
- The charge on any pixel is proportional to the number of photons absorbed.

> This means that the probability of an individual photon liberating an electron is 0.7.

Information from each pixel is read and interpreted by a computer. The charge-coupled device has a much higher quantum efficiency than photographic film, at more than 70% compared to 4%.

Radio astronomy

AQA A   M6A

> Using wire mesh reduces the weight of the dish and the size of the structures needed to support it.

A **radio telescope** has a similar structure to that of a reflecting optical telescope, but they need to be much larger for resolution of the longer wavelengths involved. In a radio telescope:

- the reflector is a parabolic dish which can be made from mesh provided that the mesh size is less than $\lambda/20$, where λ is the wavelength being detected
- the radio waves are detected by an aerial placed at the principal focus of the reflector
- the reflector should be manufactured to a precision of $\lambda/20$ to avoid distortion
- the resolving power is the same as that for an optical telescope, $\theta \approx \dfrac{\lambda}{D}$,
- doubling the diameter of a dish quadruples its area and the power received, so $P \propto d^2$
- the reflector is movable so that an individual object can be tracked or a specific area of the sky can be scanned.

Progress check

1. What is the advantage of a reflecting optical telescope over a refracting one?
2. A radio telescope has a dish diameter of 75 m. What is its resolving power when detecting 0.50 m wavelength radio waves?

1 It can be made of much larger diameter, allowing increased resolution and brighter images.

2 6.7×10^{-3} radians

6.2 The model unfolds

After studying this section you should be able to:

- describe how the work of Copernicus, Kepler and Galileo enabled greater understanding of the Solar System and the Universe
- apply the three methods of measuring astronomical distances in appropriate situations
- explain how the evolution of a star depends on its mass

LEARNING SUMMARY

Early models

OCR A ▷ M5.1

Copernicus' model of the Universe (circa 1520), in which he proposed that the Sun and not the Earth, is at the centre, was the first model that explained the apparent backwards motion and the change in brightness of the planets. The stars were pictured as being fixed on a sphere beyond the planets.

Kepler later took the model one stage further (circa 1600) and established that the planets move in elliptical orbits. Kepler's three laws of planetary motion are:

1 planets follow elliptical orbits around the Sun, with the Sun at one focus

2 a line joining a planet to the Sun sweeps out equal areas in equal times

3 $T^2 \propto R^3$, where T is the orbital period of a planet and R is its mean distance from the Sun.

> Unlike Copernicus, Galileo did publish his findings. His support for the Copernican model lead to his imprisonment. The last years of his life were spent under house arrest.

In order to avoid conflict with the established teachings of the Church, Copernicus did not publish his ideas until just before his death in 1543. When **Galileo** (circa 1610) observed moons orbiting Jupiter, he realised that this confirmed the Copernican model and that the Church's teaching that 'everything in the Universe orbits the Earth' had to be wrong.

Having established his laws of motion on Earth, **Newton** applied these to the movement of the planets (circa 1690). This lead to the realisation that for Kepler's second law to be true there had to be a force directed inwards. He described that force in his law of universal gravitation (see section 3.1).

> **KEY POINT**
>
> Newton's law of gravitation can be used to show the relationship between the orbital period, T, and radius of orbit, R:
> $$T^2 = \left(\frac{4\pi^2}{GM_s}\right)R^3$$
> which is in agreement with Kepler's third law.

> Kepler's law was derived by analysing data, so it provides experimental support for Newton's law.

Although Newton's law and Kepler's laws supported each other, the orbit of Uranus did not. Uranus was seen to change orbital speed in a way that Newton's theory did not explain. Not, that is, until the theory was used to predict the existence of Neptune as being responsible for the irregularities in Uranus' orbit.

The modern picture

OCR A ▷ M5.1

The **Solar System** includes:

- the **Sun**, the star at its centre and its source of energy
- the **planets**, which can be seen by reflected sunlight
- planetary **satellites**, the natural satellites are the planets' moons
- **comets**, which unlike the planets can orbit the Sun in any plane and any direction.

The **Universe** includes:

- galaxies; a **galaxy** is a collection of stars, between a few million and hundreds of billions in number, held together by gravitational forces
- an uncountable number of stars
- **microwave radiation**, thought to be left over from the creation of the Universe, and corresponding to a temperature of 3 K.

Astronomical distances

AQA A M6A OCR A M5.1

The AU is used for measuring distances within the Solar System, the light year for close stars and the parsec for more distant stars.

Different units are used to measure distances in our Solar System, our galaxy and beyond our galaxy. The table shows these units and gives their equivalent in m.

unit	definition	equivalent distance in m
astronomical unit (AU)	mean distance between Earth and Sun	1.5×10^{11}
light year	distance travelled by light in 1 year	9.5×10^{15}
parsec (pc)	distance between observations in AU / angle subtended (in arc seconds, where 1 arc second = 1/3600 degrees)	3.1×10^{16}

The parsec is used to measure distances by observing the position of a star from opposite extremes of the Earth's orbit. At these extremes the distance between observations is 2 AU.

The next table shows typical sizes and masses of the Universe and its contents.

object	distance in light years	mass in kg
nearest star	4.3	10^{30}
Milky Way galaxy	10^5 measured across from opposite extremes	10^{40}
Universe	10^{10} measured across from opposite extremes	unknown, but thought to be around 10^{55}

The life of a star

AQA A M6A OCR A M5.1

When protons fuse together the resulting nucleus has less mass than its constituent particles. Einstein's equation $E = mc^2$ is used to calculate the energy released in fusion (see section 4.4).

When a star is formed:

- clouds containing dust, hydrogen and helium collapse due to gravitational forces
- the contraction causes heating
- the temperature in the core becomes hot enough for fusion reactions to occur
- energy is released as hydrogen nuclei fuse, eventually forming the nuclei of helium.

The star is now in its **main sequence**. How long it remains like this depends on its mass; the more massive the star, the shorter the main sequence.

When there is no longer sufficient hydrogen in the core of a star to maintain its energy production and brightness the star cools:

Fusion to form nuclei more massive than iron requires energy to be supplied, so energy cannot be released by further fusion reactions when the core consists of iron.

- the core contracts but the outer layers expand, forming a **red giant** or **red supergiant**, depending on its size
- increased heating of the core creates temperatures hot enough for the helium nuclei to fuse, forming carbon and oxygen.

In a small star like our Sun, once the helium has fused the core contracts to form a **white dwarf**, and the outer layers are ejected. The star then cools to an invisible **black dwarf**.

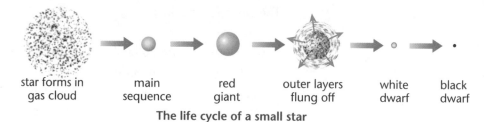

| star forms in gas cloud | main sequence | red giant | outer layers flung off | white dwarf | black dwarf |

The life cycle of a small star

In a larger star:

- further fusion reactions occur and the star becomes a blue supergiant
- the nuclei of more massive elements are formed until the core is mainly iron, the most stable element
- the core now collapses and the heating causes the star to increase in brightness, it is a supernova, an exploding star
- in the explosion the outer layers are flung off to form a new dust cloud and the dense core remains as a neutron star or, in the case of a very massive star, a black hole, whose gravitational field is so strong that light cannot escape from it.

The boundary around a black hole, beyond which nothing can be seen, is called the event horizon. Its size is related to the mass of the black hole.

> **KEY POINT**
>
> The **Schwarzschild radius**, R_s is the radius of the event horizon around a black hole:
>
> $$R_s = \frac{2GM}{c^2}$$
>
> where M is the mass of the black hole, G is the universal gravitational constant and c is the speed of light.

Stars and galaxies

OCR A   MS.1

The Sun is a small star situated on one arm close to the outer edge of a spiral galaxy called the Milky Way. The diagram shows the three shapes of galaxy: spiral, elliptical and irregular.

The Milky Way galaxy whirls around at enormous speeds but, like the planets in the Solar System, the orbital time of stars increases with increasing distance from the centre. When Newton's law of gravitation is applied to the Solar System, the mass of the planets is ignored as this has only a small effect on the movement of an individual planet. The mass of all the other stars in a galaxy cannot be ignored when applying the law to an individual star, but only those within the orbit of a star are taken into account as it is the gravitational attraction of these stars that provides the centripetal force.

> **KEY POINT**
>
> Newton's law applied to the orbit of a star gives the relationship between the orbital time, T, and the orbital radius, r:
>
> $$T^2 = \frac{4\pi^2 r^3}{GM}$$
>
> where M is the total mass contained within the orbit of the star.

Olber's paradox

OCR A   M5.1

A paradox is an apparent contradiction.

Newton pictured a static Universe that stretched to infinity, with stars being uniformly distributed throughout. Otherwise, he argued, the Universe would collapse due to the gravitational attraction of its mass. Olber's paradox is that if this were the case the sky should be bright at night. To understand why:

- imagine that the Universe is split into shells of equal thickness, like the structure of an onion
- the intensity of the light received from a star in a shell of radius r is four times that received from a star in a shell of radius $2r$

The volume of a shell of surface area A and radius r is approximately equal to Ar.

- because it has four times the volume, a shell of radius $2r$ contains four times as many stars as one of radius r
- it follows that the light from each shell is equally bright, and if there are an infinite number of shells the sky should be bright at night.

The contradiction arises because of the incorrect assumptions that:

The Universe is finite and expanding.

- the Universe is infinite
- the Universe is static.

The cosmological principle is a statement that there is nothing special about the Earth. It states that:

- on a large scale, mass is distributed evenly throughout the Universe, although on a small scale there are regions of high and low concentration
- on a large scale, the Universe does not differ in any one direction compared to any other direction
- physical laws and principles apply equally throughout the Universe.

Progress check

1 The second nearest star to Earth subtends an angle of 1.52 arc seconds when measured from opposite extremes of the Earth's orbit. Calculate its distance from the Earth in pc.

2 What causes the release of energy from a star in its main sequence?

3 What is meant by an *event horizon*?

3 The boundary around a black hole, beyond which nothing can be seen.
2 The fusion of protons into more massive nuclei, eventually forming helium.
1 1.32 pc.

6.3 The expanding Universe

After studying this section you should be able to:

- *explain how line spectra give evidence for the composition of stars*
- *describe how Doppler shift can be used to measure the speed of a star relative to the Earth*
- *explain how Hubble's law can be used to estimate the age of the Universe*

LEARNING SUMMARY

Emission and absorption spectra

In section 2.5 it was explained how line spectra are emitted when an electric current passes in an ionised gas. Light is emitted when an electron loses energy either by being captured by a gas ion or by a transition to a lower energy level.

A similar effect causes a gas that is hot (typically in excess of 1000°C) to emit light. When this happens:

- electrons gain energy in the collisions due to thermal motion of the particles
- electrons can only absorb the amounts of energy associated with movement to a higher energy level
- electromagnetic radiation is emitted when an electron loses energy and moves to a lower energy level.

> The temperature at which a gas starts to emit light depends on the smallest energy level transition that results in the emission of a visible photon.

As each element has its own characteristic emission spectrum, the composition of a gaseous object that emits its own light can be determined by analysing this spectrum.

This is how helium was discovered in solar flares, but the spectrum obtained from the body of the Sun and other stars is not a line emission spectrum but a line absorption spectrum. It contains the full range of visible wavelengths with a large number of 'black' lines, where wavelengths are missing.

Inside a star the particles are very close together, and their energy levels overlap. So unlike a gas at normal temperatures and pressures on Earth, there are no limits on the energy level transitions that are possible. This explains why a star emits the full frequency range of the visible spectrum. The explanation of the black lines is:

> Hot solids and liquids can emit all wavelengths, and appear white hot, because there are no restrictions on the allowed energy level transitions.

- a gas can only absorb the wavelengths of electromagnetic radiation that it emits, since these correspond to the same energy level transitions in the opposite directions
- when white light from the body of the Sun passes through the cooler, gaseous surface layers, the elements present absorb their own characteristic wavelengths
- this energy is re-emitted in all directions, much of it back into the body of the Sun, leaving a dark line in the spectrum that reaches the Earth.

The composition of a star is determined from the absorption spectrum and the wavelengths missing from that spectrum, rather than the wavelengths it contains.

Doppler shift

AQA A	M6A	EDEXCEL B	M5
AQA B	AS	OCR A	M5.1
EDEXCEL A	M4	OCR B	M4

Radar guns, used to measure the speed of a moving car, also use the Doppler effect.

You can tell whether an unseen aircraft is moving towards or away from you by the pitch of the engines. You receive higher frequency sounds when it is moving towards you than when it is moving away. This is due to the Doppler effect, the difference between the frequency of a wave emitted and that of the wave received when there is relative movement of the wave source and an observer.

> **KEY POINT**
>
> For light, the change in frequency, Δf, or wavelength, $\Delta \lambda$, due to relative motion is given by:
> $$\Delta f/f = \Delta \lambda/\lambda = v/c$$
> where v is the speed of the source relative to the observer and c is the speed of light.

The spectral lines in the absorption spectrum of a star observed from the Earth are subject to Doppler shift:

- a shift towards the blue end of the spectrum corresponds to a reduction in wavelength and an increase in frequency; this happens when a star is moving towards the Earth
- a shift towards the red end of the spectrum corresponds to an increase in wavelength and a reduction in frequency.

Hubble and red-shift

AQA A	M6A	OCR A	M5.1
AQA B	AS	OCR B	M4
EDEXCEL A	M4		

We live in an expanding Universe. Evidence for this comes from the shift in frequency and wavelength of the light received from other galaxies. Light from the nearby Andromeda galaxy is shifted towards the blue end of the spectrum, but for every other galaxy the shift is towards the red. The amount of shift depends on the speed of the galaxy relative to the Earth, known as the speed of recession.

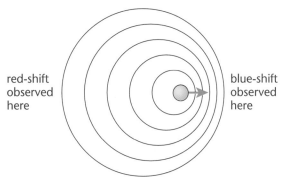

red-shift observed here

blue-shift observed here

How movement of a star causes a difference between the wavelengths emitted and those received

Hubble discovered that:

> the speed of recession of a galaxy is proportional to its distance from the Earth.

This is known as Hubble's law, and it gives rise to Hubble's constant.

Hubble's law seems to state that the Earth is the centre of the expansion. This is not the case. Measurements taken from any point in an expanding Universe would give the same pattern.

> **KEY POINT**
>
> Hubble's law can be written as
> $$v \propto d \text{ or } v = Hd$$
> where v is the speed of a galaxy relative to the Earth, d is its distance from the Earth and H is Hubble's constant.
> H has units s^{-1} but its value is not known precisely.
> An alternative unit for H is km s^{-1} Mpc^{-1}.

The expansion of the Universe gives evidence to support the Big Bang theory. According to this theory:

- the Universe started with an enormous explosion about 15 thousand million years ago
- initially the Universe was very small and very hot
- the Universe has been cooling as the galaxies have been moving away from each other ever since.

The theory is also supported by the existence of microwave background radiation that fills the whole Universe. This is thought to be electromagnetic radiation left over from the initial explosion. Due to the expansion of the Universe its wavelength has increased and its energy has decreased.

> The energy of the background radiation is equivalent to that radiated from an object at a temperature of about 3 K.

The age of the Universe

AQA A   M6A   OCR A   M5.1
EDEXCEL A   M4

Measurements of the value of the Hubble constant enable the age of the Universe to be estimated. This involves measuring both the relative speed and the distance of a star from the Earth. To estimate the distance to a star:

- for close stars (up to 300 light years) the apparent movement of the star across the fixed background of more distant stars is used. This involves measuring the position of the star when the Earth is at opposite points in its orbit.
- for more distant stars observations of the spectrum, brightness and periodic variation in brightness are needed.

> The age of the Universe is simply $\frac{1}{H}$.

Both methods only produce an estimate. In the case of close stars, movement of the star itself and the Sun also affect the apparent change in position due to the Earth's movement. For more distant stars, scattering of light from the star as it passes through the Earth's atmosphere leads to uncertainty in the measurements.

Since the distance of a star, d, cannot be measured very precisely it follows that there is uncertainty about the value of the Hubble constant, H. Probably the only certain thing about it is that it is not constant! Its value is probably decreasing due to galaxies slowing down because of the effect of their gravitational pulls on each other.

> The most recent data from the Hubble telescope puts the value of the Hubble constant at between 1.95×10^{-18} s^{-1} and 2.28×10^{-18} s^{-1}.

The diagram shows how the speed of a galaxy relative to the Earth varies with distance.

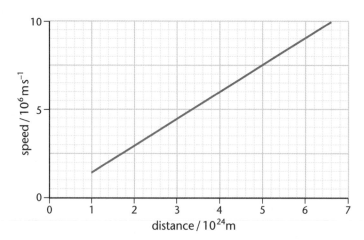

The Hubble constant can be calculated from any pair of values obtained from the graph. An estimate of the age of the Universe is then obtained by calculating the reciprocal of this constant.

In the beginning

AQA A ▸ M6A OCR A ▸ M5.1

There is little direct evidence to support any theory about what happened in the first fraction of a second after the 'Big Bang'. The energies of the particles that existed would be so vast that it is impossible at the present time to accelerate particles to these energies in order to study their behaviour.

Before the first 0.01 s:

- electrons and positrons (anti-electrons) and quarks existed
- quarks combined to form nucleons and their anti-particles.

After the first 0.01 s.

> An anti-particle has the same mass as its corresponding particle, but the opposite charge (if any). When they collide they cease to exist, forming gamma rays or other particles.

- annihilation of nucleons and anti-nucleons occurred, leaving an excess of nucleons
- after about 15 s, electron-positron annihilation left an excess of electrons
- after about 100 s, fusion of protons formed light nuclei such as helium
- after about 5×10^5 years, atoms formed from nuclei and electrons
- stars and galaxies began to form as there was less electrostatic repulsion between particles to oppose the gravitational attraction

> The formation of neutral atoms resulted in less electrostatic repulsion.

- after about 5×10^6 years the formation of clusters of galaxies caused local variations in the density of the Universe.

What about the future?

EDEXCEL A ▸ M4 OCR A ▸ M5 1
EDEXCEL B ▸ M5

There are three possibilities for the future of the Universe:

- it will continue to expand
- it will reach a steady size
- it will contract, ending in a 'Big Crunch'.

Whatever the answer, it depends on the gravitational attraction between the galaxies, and this in turn depends on the average density, known as the mass-energy density to emphasise that the energy in the Universe has considerable mass and contributes to its density.

The future of the Universe depends on its density. The critical density is that required to hold the Universe in a steady state; if the density is less than this then the Universe will continue to expand. If the density is greater than the critical density, the Universe will stop expanding and contract, resulting in the 'Big Crunch'.

Newton's law of gravitation, together with Hubble's law, can be used to show that:

> A strict derivation of the relationship is only possible using relativity theory (see section 7.5).

KEY POINT

The critical density, ρ_0, is given by the expression

$$\rho_0 = 3H^2/8\pi G$$

This gives a value of approximately 2.0×10^{-26} kg m^{-3}, depending on the value assumed for Hubble's constant.

The following graphs show the variation in the size of the Universe for the open Universe (infinite expansion) and closed Universe (final contraction) models.

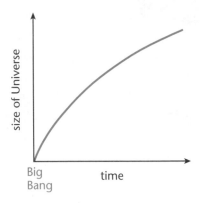 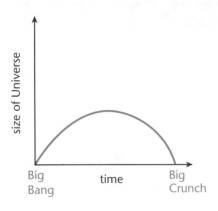

Progress check

1 Explain why dark lines in an absorption spectrum correspond to bright lines in an emission spectrum of the same element.

2 A star is moving away from the Earth at a speed of 1.0×10^6 m s^{-1}.

   Calculate the fractional difference in wavelength, $\Delta\lambda/\lambda$ between the waves received and those emitted.

3 Use data from the graph on page 145 to calculate a value for the Hubble constant.

3 1.5×10^{-18} s^{-1}

2 0.0033

1 A gaseous element can only emit and absorb certain frequencies or wavelengths. These frequencies or wavelengths are the same for emission and absorption. When white light passes through the gas it absorbs the frequencies in its emission spectrum, leaving dark lines.

6.4 Classifying stars

After studying this section you should be able to:

- *compare the apparent magnitudes and calculate absolute magnitudes of stars*
- *describe how stars are classified in terms of their brightness, temperature and composition*
- *explain how an absorption spectrum gives information about the elements present in a star*

LEARNING SUMMARY

Star magnitudes

> The magnitude scale is a logarithmic scale, equal changes on this scale correspond to the brightness changing by the same factor.

The brightness of the light received from a star, known as its apparent magnitude, depends on both its actual brightness and its distance from the Earth. A star of magnitude 1 is defined as appearing 100 times as bright as a star of magnitude 6, so one unit on the magnitude scale corresponds to a change in brightness by a factor of 2.5.

> **KEY POINT**
>
> The apparent magnitude, m, of a star is related to its intensity, I, by the expression:
> $$m = -2.5 \log I + \text{constant}$$

In using this relationship:

- the value of the constant depends on both the distance of the star from the Earth and its surface area

- since the equation is only used to compare apparent magnitudes, the constant has no significance as it does not appear in the relationship $m_1 - m_2$, the difference in the apparent magnitudes of two stars

- the ratio of intensities of two stars of apparent magnitudes m_1 and m_2 is equal to $m_1 - m_2 = 2.5^{m_1 - m_2}$.

> Apparent magnitudes are compared by the intensities of the images on photographic film or other detector exposed for a fixed time.

The smaller the apparent magnitude of a star, the brighter it appears to be. Stars can have negative values of apparent magnitude; the brightest star seen from the Earth is the Sun, with an apparent magnitude of −27.

To compare actual stellar magnitudes the effect of distance on the detected brightness needs to be taken into account. This is done by calculating the magnitudes that stars would have if they were the same distance from the Earth.

> **KEY POINT**
>
> The **absolute magnitude**, M, of a star is the apparent magnitude it would have if it were a distance of 10 pc from the Earth.
> It is calculated from the apparent magnitude using the expression
> $$m - M = 5 \log (d/10)$$
> where d is the distance of the star from the Earth in pc.

When the magnitudes of stars are plotted against their surface temperatures, the result is a Hertzsprung-Russell diagram.

The H-R diagram shows that:

- different types of star occupy specific regions on the diagram

- most stars are in the main sequence band, with the more massive stars in the upper left corner

- the white dwarfs are hot but dim because of their small size.

Note that, on the diagram, movement towards the right corresponds to decreasing temperature.

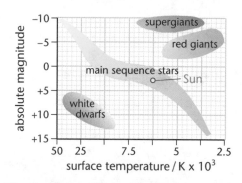

Stars are black bodies

AQA A   M6A   OCR A   M5.1

All objects emit electromagnetic radiation over a range of wavelengths that depends on the temperature of the object. Increasing the temperature of an object results in:

- an increase in the power radiated
- the wavelength at which the maximum power is radiated becoming shorter.

A black body absorbs all wavelengths of electromagnetic radiation and reflects none. However, it need not appear black, as solid objects at a temperature greater than about 600°C emit visible radiation. The rate of emission of radiation from a black body at different temperatures is shown in the graph opposite.

The area between the graph line and the wavelength axis represents the total power radiated. The graph shows that this increases with increasing temperature.

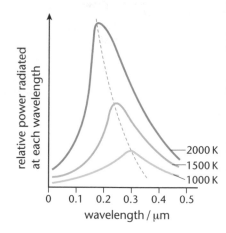

The dotted line shows how the wavelength at which the maximum power is emitted depends on the temperature of the black body. Knowledge of this wavelength enables the surface temperature of a star to be estimated using Wien's law.

Note that the unit 'm K' is 'metre × Kelvin', not milliKelvin, which would be written as mK.

> **KEY POINT**
>
> Wien's law relates the wavelength at which the maximum power is emitted by a black body, λ_{max}, to its temperature:
> $$\lambda_{max}T = 2.90 \times 10^{-3} \text{ m K}$$
> Where T is the temperature of the object in K.

In practice it may be difficult to determine λ precisely because:

- the instrument used to detect the radiation may not have the same sensitivity for all wavelengths
- the atmosphere absorbs some wavelengths of light more than others, so the intensity of these wavelengths may be reduced by more than that of others as the light travels through the atmosphere.

The power radiated by a black body depends on both its surface area and its temperature.

When applying Stefan's law to stars, it is assumed that stars are black bodies in that they are unselective in the wavelengths of radiation that they emit and absorb.

> **KEY POINT**
>
> **Stefan's law** states that:
>
> the power radiated per m² of a black body is proportional to its (absolute temperature)⁴
> $$P = \sigma A T^4$$
> where σ is Stefan's constant and has the value 5.67×10^{-8} W m⁻² K⁻⁴.

When applying Stefan's law:

> All the radiation from a star passes through a spherical surface of radius *d* with uniform intensity if the centre of the surface is at the star. Doubling the radius of the sphere causes its surface area to quadruple, so the power incident on unit area, the intensity, is quartered.

- the power radiated is calculated from measurements of the intensity of the light received and the distance of the star from the Earth, using the inverse square law

- the inverse square law states that the intensity decreases as the inverse of the square of the distance, $I \propto \dfrac{1}{d^2}$

- the inverse square law applies to light or other electromagnetic radiation travelling in a vacuum, but absorption and scattering of light by the atmosphere reduces the reliability of calculations based on this law.

If the temperature of a star is calculated from Wien's law, then Stefan's law can be used to estimate the area needed for the star to have the same power output as the Sun (3.8×10^{26} W). The surface area, and hence the radius of the star, can be calculated by comparing the absolute magnitudes.

Spectral classes

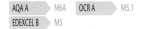

Analysis of the emission and absorption spectra of a star enables it to be classified according to its temperature and composition. The table shows the characteristics of the seven classes of star in order of decreasing surface temperature.

type	colour	surface temp/K $\times 10^3$	absorption of hydrogen Balmer lines	elements detected in absorption spectrum
O	blue	40	weak	helium ions
B	pale blue	20	fairly strong	helium atoms
A	white	10	strong	hydrogen
F	pale yellow	7	medium	ionised metals such as calcium and iron
G	yellow	6	weak	calcium ions and iron atoms
K	orange	5	weak	calcium and iron atoms, together with some molecules
M	red	3	weak	molecules of metal oxides

This table shows that:

- in the surface layers, ionised helium is only present in the hottest stars where the temperature is high enough to ionise helium atoms

- stars in groups B and A show strong absorption by hydrogen in the Balmer series; this is the part of the hydrogen emission spectrum due to exciting electrons from the first excited energy level above the ground state (see section 2.5). This energy level is given the value $n = 2$, where $n = 1$ corresponds to the ground state. Absorption of these spectral lines shows that the temperature is cool enough for hydrogen atoms to exist but hot enough for them to be in an excited state

- the relatively cool stars in classes F and G show absorption lines due to ionised metals

- stars in classes K and M are cool enough for molecules such as those of metal oxides to exist without being separated into atoms or ionised.

Analysing an absorption spectrum obtained at the Earth's surface can be misleading, as the following diagram shows.

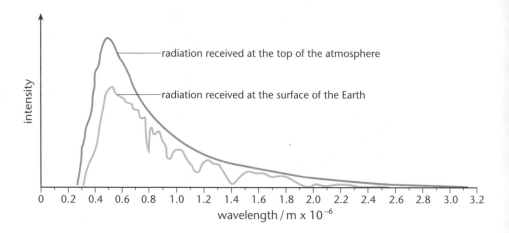

This graph shows the reduction in intensity of all wavelengths as the radiation passes through the atmosphere, with particular wavelengths being absorbed more than others. Information from spectra obtained from satellites gives a fuller picture than that from spectra obtained at ground level.

Quasars

AQA A ▷ M6A

Quasars are the most distant objects that can be detected. They are very bright objects which were first discovered by detection of the radio signals that some of them emit. Analysis of the light from a quasar reveals that:

- the spectrum is an emission spectrum, with very bright lines on a continuous spectrum
- these lines show enormous red-shift, indicating that quasars are moving with very high speeds.

Analysis of the red-shift indicates that some quasars are moving at speeds greater than 90% of the speed of light at distances of more than ten billion (1×10^{10}) light years. They are relatively small and yet they appear to emit thousands of times as much energy as a typical galaxy.

However, this may be misleading. If some other mechanism is responsible for the red-shift, they could turn out to be much closer and less bright than was at first believed.

Progress check

1 The apparent magnitude of Venus is –4, and that of the Sun is –27. By what factor does the Sun appear brighter than Venus?

2 Suggest why the magnitudes of red giants are greater than some main sequence stars that have a higher surface temperature.

3 Explain why dark lines in an absorption spectrum correspond to bright lines in the emission spectrum of the same element.

3 A gaseous element can only emit and absorb certain frequencies or wavelengths. These frequencies or wavelengths are the same for emission and absorption. When white light passes through the gas it absorbs the frequencies in its emission spectrum, leaving dark lines.

2 The red giants have a much greater surface area.

1 1.4×10^{9}

6.5 Relativity

After studying this section you should be able to:

- recall the postulates of special relativity
- calculate the amount of time dilation and length contraction when a moving object is observed
- explain why there is a maximum speed to which an object can be accelerated

LEARNING SUMMARY

Does time pass at a constant rate?

OCR A   M5.1

In everyday life, we sense and judge speed relative to the surface of the Earth or vehicles that we travel in. If you are in a car travelling at 25 m s^{-1} and another car travelling at the same speed is approaching you, then its speed relative to your vehicle is 50 m s^{-1}. If, on the other hand, it is in front of you and travelling in the same direction, its speed relative to you is zero.

This led Einstein to use the idea of a **frame of reference** as something that speeds are measured relative to. In **special relativity** frames of reference are not accelerating.

> Since there is no such thing as absolute motion, the laws of physics have to be the same for all observers in relative motion, provided they are not accelerating.

> **KEY POINT**
>
> The laws of physics apply equally in all inertial frames of reference.
>
> This is the first postulate (assumption) of Einstein's theory of special relativity.

In the diagram, the driver of the red car sees the blue car approaching at 50 m s^{-1} but, from the frame of reference of the stationary observer, the blue car is approaching at 30 m s^{-1}.

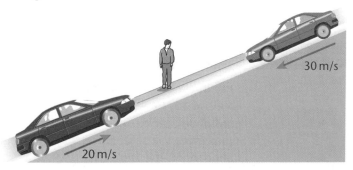

What about the light that passes between the two vehicles? Is its speed affected by their motion? The answer is no.

> **KEY POINT**
>
> The speed of light in a vacuum always has the same value.
>
> This is the second postulate of Einstein's theory.

> Far more muons reach the surface of the Earth than is predicted by a half-life of 2 microseconds.

The invariance in the speed of light leads to the conclusion that neither time nor distance have absolute values. Each of the drivers of the cars in the diagram would notice that time passes more slowly in the other vehicle. This effect is known as **time dilation**. It increases the apparent half-life of muons created in the upper atmosphere from 2 microseconds to 20 microseconds, measured by an observer on Earth. We do not normally notice the effects of time dilation because we move at speeds very much less than the speed of light. The muons move through the atmosphere at a speed which is within 1% of that of light.

As well as having different measurements of time, the car drivers in the diagram have different measurements of distance. Each driver sees the other car as being shorter than its driver sees it; this is known as length contraction. Length contraction only occurs in the direction of motion.

The factor by which time is dilated and length is contracted is $\sqrt{1-\dfrac{v^2}{c^2}}$, where v is the speed of the object relative to the observer and c is the speed of light.

It follows that time t_0 and length l_0 measured within a frame of reference are observed by someone moving relative to that frame to have values:

- $t = t_0 / \sqrt{(1 - v^2/c^2)}$
- $l = l_0\sqrt{(1 - v^2/c^2)}$

How is mass affected?

OCR A   M5.1

> For everyday objects moving at everyday speeds, there is no significant increase in mass.

A consequence of length contraction is that as an object speeds up under the action of a constant force, its apparent acceleration decreases. This causes an apparent increase in mass.

The apparent mass, m, is given by the expression

$$m = \frac{m_0}{\sqrt{(1 - v^2/c^2)}}$$

where m_0 is the **rest mass**, the mass of the object measured within its frame of reference.

This expression shows that the mass of an object tends towards infinity as its speed approaches the speed of light. It is therefore impossible to accelerate an object beyond the speed of light – the maximum possible speed.

General relativity

OCR A   M5.1

> This is known as the principle of equivalence of inertial and gravitational forces.

> The path of the light appears to be curved because of the movement of the spacecraft since the light was emitted by the torch.

Einstein stated that gravitational forces are equivalent to other forces that cause acceleration. Imagine an astronaut in a spacecraft, away from the effects of any gravitational fields. If the craft accelerates at a rate of 9.8 m s⁻², the astronaut experiences the same force that she would if she were in the capsule and it was stationary on the Earth.

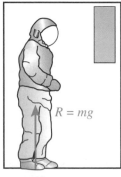

astronaut
stationary
on Earth

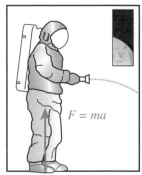

astronaut
accelerating
in space

Furthermore, light from her torch would appear to follow a curved path. So if light follows a curved path within an accelerating frame of reference, it should also follow a curved path within a gravitational field.

This was confirmed in the solar eclipse of 1991, where two stars appeared to move further apart when the Sun and Moon were between the paths of light coming from the stars. The apparent change in position was due to the light changing direction as it passed through the Sun's gravitational field.

Gravitational fields also cause time dilation, with clocks running at different speeds in gravitational fields of different strengths.

Einstein's theory of general relativity was used to explain the changes observed in Mercury's orbit, which Newton's laws could not explain. Mercury's perihelion, the point of closest approach to the Sun, rotates around the Sun. Calculations based on the general theory of relativity predicted the amount of rotation of the perihelion to be the same as that observed.

Progress check

1 A muon has a half-life of 2 microseconds, measured within its frame of reference. Calculate the observed half-life when the muon moves at $0.99c$.

2 Explain why a moving spacecraft appears shorter but not narrower to an external observer.

3 Explain whether light follows a curved path as it travels past the Earth, above the height of the atmosphere.

3 Yes, due to the effect of the Earth's gravitational field.

2 Length contraction only occurs in the direction of motion.

1 14 microseconds

Sample question and model answer

(a) Explain the meaning of the *Doppler Effect* and describe, in outline, how it is used to measure the speeds of stars relative to the Earth. [4]

The Doppler Effect refers to the change in wavelength and frequency when a source of waves is moving relative to an observer (1 mark). The wavelength is increased when the source is moving away from the observer and a decrease in wavelength when it moves towards the observer (1 mark).

An outline description means that you should describe the physical principles without going into great detail of how the measurements are made.

The faster a star is moving, the greater the change in wavelength (1 mark). To measure the speed of a star, the wavelength received is compared to that emitted by a stationary source (1 mark).

b) Measurements of *red shift* often show different values for different stars in the same galaxy. Suggest a reason for this.

The effect of the rotation of stars within a galaxy is to broaden the spectral lines received from the galaxy as a whole, showing that stars are moving with a range of speeds relative to the Earth.

Stars within a galaxy are rotating around the centre of the galaxy (1 mark). Depending on the direction of movement of a particular star, this can cause an increase or a decrease in its velocity relative to the Earth and consequently an increase or decrease in the amount of red shift (1 mark).

(c) In the light from a star, the wavelength of a hydrogen line in its absorption spectrum is measured to be 5.153×10^{-7} m. A laboratory measurement of the same spectral line gives its wavelength as 4.861×10^{-7} m.

(i) Calculate the speed of movement of the star relative to the Earth.
$c = 3.00 \times 10^8$ m s^{-1} [2]

There is no mark for the recall of this relationship as it will be given in the question or on a separate data sheet.

$\Delta\lambda/\lambda = v/c$
$v = c \times \Delta\lambda/\lambda = 3.00 \times 10^8$ m s^{-1} $\times$ $(5.153 \times 10^{-7} - 4.861 \times 10^{-7}$ m$)/$ 4.861×10^{-7} m (1 mark) $= 1.80 \times 10^7$ m s^{-1} (1 mark)

(ii) Explain whether the star is moving towards or away from the Earth. [2]

The cue word explain here means that you must give a reason to justify your answer to gain full marks.

The star is moving away from the Earth (1 mark) as the wavelength of light received is greater than that emitted (1 mark).

Practice examination questions

1

(a) (i) Explain the meaning of the terms *apparent magnitude* and *absolute magnitude*. [2]

   (ii) Explain why the absolute magnitude is preferred to the apparent magnitude when comparing the magnitudes of stars. [2]

(b) Sirius is at a distance of 2.7 pc from the Earth. Its apparent magnitude is –1.5. Calculate the absolute magnitude of Sirius. [2]

(c) The absolute magnitude of the Pole star is –4.9. Its apparent magnitude is 2.0. Calculate the distance of the Pole Star from the Earth. [2]

(d) White dwarfs are hotter than red giants, but they are less bright. Suggest a reason for this. [2]

2

The hydrogen absorption spectrum contains a dark line corresponding to a wavelength of 6.563×10^{-7} m.

(a) Explain why the lines in an absorption spectrum are dark. [3]

(b) In the spectrum detected from a star, this line has a wavelength of 6.840×10^{-7} m.

   (i) Explain this apparent increase in wavelength. [2]

   (ii) Calculate the speed of the star relative to the Earth.
   $c = 3.00 \times 10^8$ m s^{-1}. [3]

   (iii) The value of the Hubble constant, $H_0 = (2.12 \times 10^{-18} \pm 0.07)$ s^{-1}. Calculate the range of distances from the Earth within which the star lies. [3]

3

The diagram shows three possible ways in which the size of the Universe could change in the future.

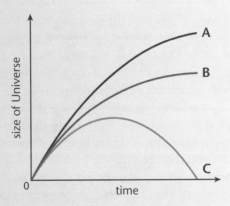

With reference to the graph, explain how the future of the Universe depends on the amount of mass that it contains. [6]

Chapter 7

Synoptic assessment

What is synoptic assessment?

Unit tests only examine the material in that unit, in isolation from any other content of the AS and A2 courses.

Synoptic assessment:

- takes place at the end of the course
- counts for 15–20% of the marks for A Level, and 30 – 40% of the marks for A2
- tests understanding, knowledge and skills in contexts that bring together different areas of the subject
- emphasises skills and principles.

What skills do I need?

All the skills that you have developed throughout your A Level course. In particular, you need to be able to:

- make links between different areas of physics – for example, how the inverse square law applies to radial fields and the spreading of light from a point source and the similarities and differences between radioactive decay and capacitor discharge
- apply knowledge and understanding in unfamiliar contexts – for example, knowledge and understanding of force and motion to vehicle safety features
- handle data – for example, use a graph to test a relationship between two quantities
- interpret and evaluate information – for example, explain the meaning of scientific terms used in a passage.

What types of question are asked?

There are a number of different types of synoptic question. The awarding bodies use a combination of these.

- Passage analysis You are given a passage which may be adapted from a book or a scientific journal. You may then be asked questions that test your comprehension of the passage, analyse data, perform calculations and draw inferences from information given in the passage.
- Structured questions These are similar to the questions in unit tests, but they will draw on content from more than one unit.
- Unstructured questions Instead of being split up into separate parts, these give you the opportunity to develop arguments and ideas. When answering these questions, you should always refer to physical principles such as conservation of momentum or electromagnetic induction.

How should I prepare for synoptic assessment?

Synoptic assessment can cover any area of the full A Level specification. It is important therefore that you:

- know all the formulae and relationships for your specification, both AS and A2, that are not provided on a data sheet
- can interpret graphs, including the significance of gradients and areas
- revise and review all the key concepts in the full AS and A2 specification

7.1 Drawing it all together

LEARNING SUMMARY

After studying this section you should be able to:

- *describe how similar physical models can be applied to a variety of situations*
- *make comparisons between the action of a spring and a capacitor and also between electric and gravitational fields*
- *explain the similarities between radioactive decay and capacitor discharge*

Comparing a spring and a capacitor

EDEXCEL A   M6

In each case the outcome variable, extension of the spring or charge stored on the capacitor, is proportional to the input variable, force applied to the spring or voltage across the capacitor. This is shown in the table.

	input variable	outcome variable	relationship
spring	force, F	extension, x	$F = kx$
capacitor	voltage, V	charge stored, Q	$V = Q/C$

The stiffness, k, of a spring is equivalent to $1/C$ for a capacitor as:

- the greater the value of k the smaller the extension for a given force
- the greater the value of C the greater the charge stored for a given voltage.

The expressions for the energy stored in a stretched spring or a charged capacitor are also similar.

> In each case the energy stored can be related to the area between the graph line and the force or voltage axis.

KEY POINT

In each case the energy stored is equal to:

½ × input variable × outcome variable

$\frac{1}{2}Fx$ and $\frac{1}{2}VQ$

for the spring and the capacitor respectively.

Electric and gravitational fields

EDEXCEL A   M6   EDEXCEL B   M5

> This statement is true for point masses and point charges. In the case of large masses and charges, it only applies to the space beyond the mass or charge.

All radial fields show an inverse square law in the relationship between the field strength and distance from the centre of the field.

The difference between gravitational and electric fields is:

- gravitational fields can only exert attractive forces on masses
- the force exerted by an electric field on a charge can be either attractive or repulsive.

The similarities between the forces exerted and the field strengths at a distance r from the centre of the field are shown in the table.

> Remember that
> $k = 1/4\pi\varepsilon_0$

type of field	size of force on point mass/charge	field strength
electric	$F = kQq/r^2$	$E = kQ/r^2$
gravitational	$F = GMm/r^2$	$g = GM/r^2$

For these fields:

- the constant k is for an electric field is equivalent to the constant G for a gravitational field
- the field strength is defined as the force per unit mass/charge.

Capacitor discharge and radioactive decay

EDEXCEL A ▷ M6

> The rate of decay or discharge falls as the number of undecayed nuclei or charge on the capacitor falls.

Unlike radioactive decay, the discharge of a charged capacitor is not a random process, but they are both examples of exponential change. This means that:

- the rate at which the charge on the capacitor changes is proportional to the charge on the capacitor
- the rate at which the number of undecayed nuclei changes is proportional to the number present
- in equal time intervals the charge on the capacitor or the number of undecayed nuclei changes in the same ratio.

In radioactive decay the time interval used is the half-life, $t_{1/2}$. In capacitor discharge the time interval used is the time constant, $\tau = RC$.

The similarities and differences between the treatment of capacitor discharge and radioactive decay are summarised in the table.

	amount or number present	rate of change	time interval
capacitor discharge	$Q = Q_0 e^{-t/RC}$	current, $I = Q/RC$	τ = time for charge to fall to $1/e$ of its initial value
radioactive decay	$N = N_0 e^{-\lambda t}$	activity, $A = \lambda N$	$t_{1/2}$ = time for number of undecayed nuclei to fall to $\frac{1}{2}$ of its initial value

The decay constant λ in radioactive decay is equivalent to $1/\tau$ in capacitor discharge. This is because:

- the larger the value of λ the greater the rate of decay
- the larger the value of τ the smaller the rate of discharge.

> $t_{1/2}$ for a capacitor is the time it takes for the charge to fall to half its initial value.

However, there is no reason why different mathematical treatments should be applied to these situations. If the concept of half-life, or half-charge, is applied to capacitor discharge another similar relationship emerges:

$$t_{1/2}/\tau = \ln 2 \quad \text{and} \quad \lambda t_{1/2} = \ln 2$$

This again reveals the equivalence of λ and $1/\tau$.

Progress check

1 **a** If two identical springs of stiffness k are joined in series, what is the stiffness of the combination?

   b If two identical capacitors of capacitance C are joined in series, what is the capacitance of the combination?

2 Give one way in which the force between two point masses is different from that between two positive point charges.

3 What fraction of undecayed nuclei remain in a radioactive sample after time $t = 1/\lambda$ elapses?

3 $1/e$
2 Either one is gravitational and the other is electric or one is attractive and the other is repulsive.
b $C/2$
1 **a** $k/2$

7.2 Accelerators and detectors

After studying this section you should be able to:

- *describe how a Van de Graaff generator and a linac are used to accelerate particles for firing at fixed targets*
- *explain how a cyclotron is used for colliding beam experiments*
- *explain the principles of operation of particle detectors*

LEARNING SUMMARY

Van de Graaff generator

EDEXCEL A   M6   EDEXCEL B   M4

Remember, 1 eV is the energy gained by an electron when accelerated through a potential difference of 1 V.

The Van de Graaff generator is used to accelerate particles in a straight line. It is capable of giving them energies up to several MeV. The principle of a high voltage Van de Graaff generator is shown in the diagram.

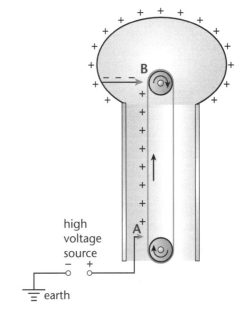

- Positive charge is sprayed onto a revolving belt at A by a high voltage source which ionises the air and repels positively-charged particles onto the belt.

Negative particles are sprayed when the surrounding gas is ionised by the strong electric field, and the negative ions are repelled by the negative electrode.

- A negative charge is induced on the inner surface of the dome, with a positive charge being induced on the outer surface.

- Negatively-charged particles are sprayed onto the belt at B.

This results in a very high voltage being created between the dome and earth. The voltage can be used to accelerate charged particles by passing them through a series of electrodes connected so that ions pass through them in stages. At each stage the ions are accelerated and focused.

The diagram below shows an arrangement suitable for accelerating negative ions such as electrons. To accelerate positive ions, the potential difference is reversed.

The resistors are used to create a 'step' in potential from each cylinder to the next.

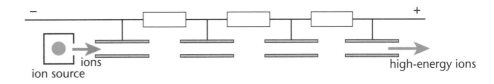

The linac

AQA B   M5   NICCEA   M6
EDEXCEL A   M6   WJEC   M5
EDEXCEL B   M4

A linac (short for linear accelerator) provides much higher-energy particles than a Van de Graaff machine is capable of producing. Charged particles accelerated by a linac can achieve energies of several GeV, at which their speed is approaching that of the speed of light.

No particle can ever reach the speed of light because, according to Einstein's theory, the mass of a particle increases with increasing speed and approaches infinity as its speed approaches that of light.

The diagram below shows a linac that is used to accelerate protons.

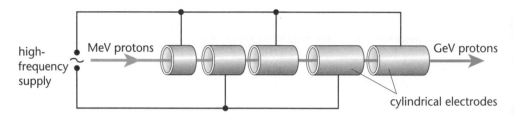

In a proton-accelerating linac:

- the protons pass through a large number of cylindrical tubes
- the protons travel at a constant speed within each tube and are accelerated as they pass into the next one
- the acceleration is achieved by a high-frequency, high-voltage alternating potential difference applied to the tubes
- the polarity of the potential difference applied to each tube changes to positive as the proton leaves the previous tube
- each tube is longer than the previous one because the accelerated protons are travelling increasing distances in the time taken for half of one complete cycle of the applied voltage to occur.

As the protons speed up, they travel increasing distances in time $1/2f$. This is the time it takes to travel through one tube and pass into the next, where f is the frequency of the voltage supply.

The source of protons is normally a Van de Graaff accelerator, so that the protons entering the linac already have energies of MeV. Linacs that accelerate electrons use high-frequency electromagnetic waves that travel inside metal cavities.

The World's largest linear accelerator is at Stanford in California. It is nearly 4 km long and it can accelerate electrons and positrons to energies of 50 GeV. Linear accelerators:

- produce a narrow beam of high intensity
- suffer little energy loss from the beam as it travels along the accelerator.

There are plans to build a new electron-positron linac that will be 30 km long and will accelerate these particles to energies of the order of 10^{13} eV.

Ring accelerators

AQA B	M5	NICCEA	M6
EDEXCEL A	M6	WJEC	M5
EDEXCEL B	M4		

Ring accelerators can produce even higher energy particles than linacs currently in use. They can accelerate particles to energies of 1000 GeV, without requiring an enormously long tube. Protons can be accelerated to speeds approaching the speed of light in a 200 m-diameter ring. The principle of the cyclotron is explained in section 3.5. In a cyclotron:

- the particles follow a spiral path within two semi-circular dees
- they are accelerated each time they cross from one dee into the other
- a magnetic field perpendicular to the dees keeps the particles in circular motion.

This is how a transmitting aerial emits radio waves; electrons emit radiation when their velocity changes.

One major disadvantage of ring accelerators is that the charged particles radiate energy due to the continual change in velocity as they change direction – a problem that does not arise in a linear accelerator.

Targets

AQA B	M5	EDEXCEL A	M6

High-energy particles from accelerators are used in the quest for more knowledge and understanding of the behaviour of sub-atomic particles. To achieve this, high-energy particles are fired at targets and then the tracks left by the collision debris are examined.

The particles produced by an accelerator can collide with a fixed target or another beam of particles in a colliding beam experiment.

A fixed target could be liquid hydrogen or water or a metal such as iridium. In a fixed target experiment:

- conservation of momentum applies to these collisions, so the momentum of the particles after the collision is equal to that of the accelerated particle that collides with the target
- the kinetic energy of the particles formed as a result of the collision comes from that of the accelerated particle
- very little energy, typically only 5%, is available to create new particles.

When beams of particles moving in opposite directions collide:

- conservation of momentum applies, so if the particles have equal masses and speeds their momentum after the collision is zero
- no kinetic energy of the colliding particles remains after the collision, so all of the initial kinetic energy is available to create new particles
- these collisions can result in the creation of a new particle that has no kinetic energy but is high in mass.

> Conservation of momentum applies to all collisions, but when the resulting particles have a large amount of momentum, they also have a large amount of kinetic energy.

> Energy has mass, and mass-energy is conserved. A loss in kinetic energy results in the mass of that energy appearing elsewhere.

Tracks

AQA B   M5   EDEXCEL A   M6

There are a number of ways of detecting the particles created in high-energy collisions.

In a cloud chamber, particles pass through a super-saturated vapour which condenses on the ionised air particles created by collisions, leaving vapour trails.

A similar effect occurs in a bubble chamber. Here, the particles pass through a superheated liquid, and bubbles of vapour form in the path of the particle. Liquid hydrogen is often used as it acts both as the fixed target and the detector.

Because the liquid in a bubble chamber is much denser than the gas-vapour mixture in a cloud chamber, collisions are more frequent and high-energy particles are stopped in a shorter distance. A magnetic field applied perpendicular to the plane of the chamber causes charged particles to follow curved paths.

The diagram below shows cloud chamber and bubble chamber tracks.

> Hydrogen is particularly suitable as a target as its nucleus is a single proton.

> The energies of the alpha particles can be deduced from the tracks that they leave.

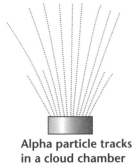

Alpha particle tracks in a cloud chamber

Bubble chamber tracks

These tracks show that:

- the alpha particles in the cloud chamber have two different values of energy and momentum
- the bubble chamber shows the tracks left by particles of opposite signs, since there are deflections in opposite directions due to the magnetic field
- there is a range of particle energy and momentum.

In a spark chamber, ionising particles cause sparks as they pass between electrodes and ionise an inert gas such as neon. The passage of particles through the chamber can be detected in two ways:

- by photographing the sparks
- by detecting the electrical pulses that pass in the electrodes.

> A spark is a flash of light emitted when positive ions capture electrons.

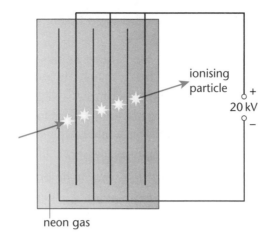

neon gas

The track of a particle through a drift chamber is produced by computer analysis of the pulses of current in the wires.

The principle of a spark chamber is shown in the diagram above.

The **drift chamber** is a development of the spark chamber and works on a similar principle. It consists of a mesh of fine wires, each of which is at the opposite potential to its neighbours. A drift chamber:

- takes into account the time taken for ionised particles to reach a wire and be detected
- can pinpoint the position of a particle to within 50 μm
- requires a very pure inert gas
- gives a three-dimensional picture of the track of a particle.

Progress check

1 Explain why the tubes in a linac are of increasing length.

2 Why is there little energy loss from particles in a linac compared to that in a ring accelerator?

3 Suggest why X-rays do not leave tracks in a cloud chamber.

3 X-rays are only weakly ionising.
2 The particles do not radiate energy due to changes in direction.
1 The time taken for the particles to pass from one tube to the next is fixed. As the particles speed up, they travel increasing distances in this time.

Sample synoptic questions and model answers

1

The gravitational field due to a point mass and the electric field due to a point positive charge show some similarities and differences.

One similarity is that both follow inverse square laws.

(a) (i) Write down expressions for the gravitational field strength due to a point mass, m, and the electric field strength due to a point positive charge, q. [2]

The definitions of gravitational and electric field strength are 'force per unit mass' and 'force per unit positive charge'.

$g = Gm/r^2$ 1 mark
$E = q/4\pi\varepsilon_0 r^2$ 1 mark

(ii) What feature of these expressions shows that the fields follow an inverse square law? [1]

All radial fields follow inverse square laws.

The r^2 term in the denominator shows that the field strengths vary as the inverse of the square of the distance from the mass or charge.   1 mark

(iii) State one other way in which the definitions of gravitational field strength and electric field strength are similar. [1]

They are both proportional to the size of the mass or charge.   1 mark

(iv) State one way in which the definitions of gravitational field strength and electric field strength are different. [2]

A negative charge experiences a force the same size as that experienced by a positive charge, but in the opposite direction.

The definition of gravitational field strength does not specify the sign of the mass as mass can only have positive values.   1 mark
The definition of electric field strength specifies that it is the size of the force on a positive charge.   1 mark

(b) Gravitational potential and electric potential are properties of points within a field.
How is potential related to the potential energy of:

(i) a mass in a gravitational field? [1]

Potential means 'the energy per unit mass/charge placed at that point'.

Potential energy = potential × mass.   1 mark

(ii) a charge within an electric field? [1]

Potential energy = potential × charge (including the sign of the charge).   1 mark

(c) Explain why the expressions for gravitational potential and electric potential have opposite signs. [4]

Gravitational forces can only be attractive (1 mark) so a point in a gravitational field always has a negative potential since energy would be needed to remove the mass to infinity (1 mark). The direction of an electric field is taken to be the direction of the force on a positive charge (1 mark). A positive charge in the field of another positive charge has a positive amount of energy since work would have to be done on the charge to move it from infinity to a point in the field (1 mark).

Sample synoptic questions and model answers (continued)

2

A motorist sees a bright light in the sky and thinks that it is the planet Venus. Later, he notices that the movement across the sky is too rapid for a planet and realises that he is looking at a low-flying aircraft.

At its brightest, Venus reflects electromagnetic radiation from the Sun with a power of 2.8×10^{17} W. On this occasion, the distance between Venus and the Earth was 8.0×10^{10} m.

The power of the lights of a large aircraft is 1500 W.

Estimate the distance of the aircraft from the motorist for its lights to appear as bright as Venus. State the assumptions that it is necessary to make and explain whether your answer is likely to be too high or too low. [12]

The marks are awarded here for setting up a workable mathematical model – like the Earth, only half of Venus is receiving radiation from the Sun at any one time.

Assuming that Venus acts as a disc (1 mark) and reflects radiation over a hemisphere (1 mark):

You are not expected to know that the surface area of a sphere = $4\pi r^2$, this information is given in a data book.

Intensity of light from Venus at the Earth
$= P/2\pi r^2$ 1 mark
$= 2.8 \times 10^{17}$ W $\div$ $(2 \times \pi \times (8.0 \times 10^{10}$ m$)^2)$ 1 mark
$= 7.0 \times 10^{-6}$ W m^{-2} 1 mark

At a distance d, for the aircraft lights to have the same intensity, assuming that they radiate light equally in all directions over a hemisphere   1 mark
    1500 W $\div$ $(2 \times \pi \times d^2)$ $= 7.0 \times 10^{-6}$ W m^{-2} 1 mark
$d = \sqrt{(1500 \text{ W} \div (2 \times \pi \times 7.0 \times 10^{-6} \text{ W m}^{-2}))}$ 1 mark
    $= 5.8 \times 10^3$ m or about 6 km   1 mark

It is important when making an estimate to round off the answer to a sensible number of significant figures.

This is assuming that the light from Venus and the aircraft contains the same proportion of visible radiation.   1 mark

The reasons for the answer are important here. No marks are awarded for a 50–50 guess.

If this is the case, then at a distance of 6 km the aircraft is likely to appear brighter than Venus, since its lights are focused into a beam, (1 mark) so I would expect my answer to be too low (1 mark).

Practice examination questions

1

A mass spectrometer is a device for comparing atomic masses. Positive ions pass through slits A and B before entering a region where there is both an electric and a magnetic field.

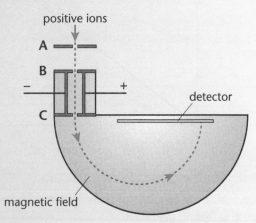

(a) Suggest why the ions must first pass through slits A and B. [1]

(b) State the direction of:
   (i)   the electric force on the ions
   (ii) the magnetic force on the ions
   (iii) the magnetic field. [3]

(c) Show that an ion can only pass through slit C if its velocity is equal to the ratio of the field strengths. [2]

(d) (i) Explain why an ion follows a circular path after leaving slit C. [2]
   (ii) An ion of carbon-12 has a mass of 2.0×10^{-26} kg and carries a single positive charge of 1.6×10^{-19} C. It emerges from slit C with a speed of 2.4×10^5 m s^{-1}.
        The magnetic field strength is 0.15 T.
        Calculate the radius of its circular path. [3]

(e) Radiocarbon dating is a method of estimating the age of objects made from materials that were once alive. In a living object the ratio of carbon-12 atoms to carbon-14 atoms is 10^{12} to one.
   Suggest how a mass spectrometer could be used to estimate the age of wood from an old table. [3]

2

In a Van de Graaff generator, charge from a moving belt is transferred onto a spherical metal dome which is insulated from the ground.

The graph below shows how the potential of a dome of radius 0.12 m depends on the charge on it.

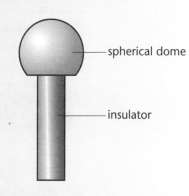

Practice examination questions (continued)

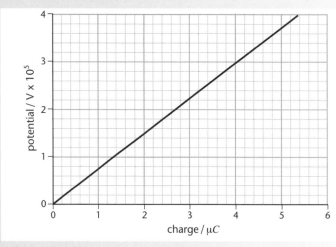

(a) Calculate the capacitance of the dome. [3]

(b) Calculate the amount of work that has to be done to place 4.0 μC of charge on the dome. [3]

The dome loses charge when the electric field strength at its surface reaches a value of 3.0×10^6 N C^{-1}.

(c) (i) Calculate the charge on the dome when the electric field strength at its surface has this value. [3]

$k = 1/4\pi\varepsilon_0 = 9.0 \times 10^9$ N m^2 C^{-2}

(ii) What is the potential of the dome when the electric field strength at its surface has this value? [1]

(d) (i) Calculate the force on an electron, charge = -1.6×10^{-19} C, near the surface of the dome when the electric field strength is 3.0×10^6 N C^{-1}. [3]

(ii) Suggest how the dome loses charge. [3]

3

Read the passage and answer the questions that follow.

Seismometers on the Moon's surface have detected very slight tremors in the body of the Moon. Some of these occur at irregular intervals and appear to originate at the Moon's surface. Others are regular and are repeated on a 14-day cycle. These regular Moonquakes originate deep below the Moon's surface and could be due to tidal stresses caused by the Moon's elliptical orbit.

It is not known whether the Earth undergoes similar tidal stresses, since the small magnitude of the tremors means that they would be indistinguishable from movement caused by other factors.

(a) Suggest one cause of the irregular tremors. [1]
(b) What 'other factors' could cause small Earth tremors? [2]
(c) Here is some data about the Moon and its orbit.

Mass of Moon = 7.4×10^{22} kg
Orbital period = 27.3 days
Closest distance between Moon and Earth = 3.56×10^8 m
Greatest distance between Moon and Earth = 4.06×10^8 m.

Estimate the difference between the greatest and smallest forces that the Earth exerts on the Moon. State clearly any assumptions and approximations that you make and what effect these are likely to have on your answer. [12]

Practice examination answers

1 Mechanics

1 (a) $\Delta E_p = mg\Delta h$ [1] = 40 kg × 10 m s^{-2} × 0.60 m [1] = 240 J [1].
   (b) $\frac{1}{2}mv^2$ = 240 J [1] $v = \sqrt{(2 \times 240\ J \div 40\ kg)}$ [1] = 3.5 m s^{-1} [1].
   (c) The child is accelerating [1] upwards [1].
   (d) $F = mv^2/r$ [1] = 40 kg × (3.5 m s^{-1})^2 ÷ 3.2 m [1] = 150 N [1].
   (e) The downward gravitational pull [1] of the Earth [1].
   (f) (i) 400 N [1] (ii) 550 N [1].

2 (a) Taking momentum to the right as positive, 0.30 kg m s^{-1} [1] – 0.16 kg m s^{-1} [1] = 0.14 kg m s^{-1} [1].
   (b) 0.14 kg m s^{-1} ÷ 1.00 kg [1] = 0.14 m s^{-1} [1] to the right [1].
   (c) Total kinetic energy before collision = 0.107 J [1].
   Kinetic energy after collision = 9.8 × 10^{-3} J [1].
   The collision is inelastic as there is less kinetic energy after the collision than before [1].

3 (a) $v = 2\pi r/T$ [1] = 2 × π × 4.24 × 10^7 m × (24 h × 60 min h^{-1} × 60 s min^{-1}) [1] = 3.08 × 10^3 m s^{-1} [1].
   (b) Acceleration = v^2/r [1] = (3.05 × 10^3 m s^{-1})^2 ÷ 4.24 × 10^7 m [1] = 0.22 m s^{-2} [1] towards the centre of the Earth [1].
   (c) The gravitational pull [1] of the Earth [1].

4 (a) 0.06 kg × 15 m s^{-1} [1] – – 0.06 kg × 24 m s^{-1} [1] = 2.34 kg m s^{-1} [1].
   (b) Force = rate of change of momentum [1] = 2.34 kg m s^{-1} [1] ÷ 0.012 s [1] = 195 N [1].
   (c) 195 N
   (d) From kinetic energy of the ball and racket [1] to potential energy of the squashed ball and stretched strings [1] and back to kinetic energy of the ball [1].

5 (a) 600 N × cos 20° [1] = 564 N [1].
   (b) Friction between the ground and the log [1]; 564 N [1].
   (c) $W = F \times s$ [1] = 564 N × 250 m [1] = 1.41 × 10^5 J [1].
   (d) It is transferred to the log and the ground [1] as heat [1].

6 (a) v^2/r [1] = (14 m s^{-1})^2 ÷ 100 m [1] = 1.96 m s^{-2} [1].
   (b) $F = ma$ [1] = 800 kg × 1.96 m s^{-2} [1] = 1.57 × 10^3 N [1].
   (c) (i) The frictional push [1] of the road on the tyres [1].
       (ii) The frictional push [1] of the tyres on the road [1].
   (d) Friction between the tyres and the road is reduced [1] and may not be sufficient for the centripetal force required [1].

7 (a) 5.0 × 10^5 kg × 60 m s^{-1} [1] = 3.0 × 10^7 kg m s^{-1} [1].
   (b) 3.0 × 10^7 kg m s^{-1} ÷ 20 s [1] = 1.5 × 10^6 N [1].
   (c) Resistive forces oppose the motion [1] so a greater forwards force is needed to produce this unbalanced force [1].
   (d) The aircraft gains momentum in the forwards direction [1]. The exhaust gases gain an equal amount of momentum in the opposite direction [1].

8 (a) $\Delta E_p = mg\Delta h$ [1] = 3.2 × 10^5 kg × 10 m s^{-2} × 180 m [1] = 5.76 × 10^8 J [1].
   (b) $\frac{1}{2}mv^2$ = 5.76 × 10^8 J [1] $v = \sqrt{(2 \times 5.76 \times 10^8\ J \div 3.2 \times 10^5\ kg)}$ [1] = 60 m s^{-1} [1].
   (c) $\frac{1}{2}$ × 3.2 × 10^5 kg × (60 m s^{-1})^2 – $\frac{1}{2}$ × 3.2 × 10^5 kg × (3 m s^{-1})^2 [1] = 5.7 × 10^8 J [1].

2 Waves

1 (a) The frequency an object vibrates at, when it is displaced from its normal position [1] and released [1].

(b) When the frequency of a forcing vibration [1] is equal to the natural frequency [1].

(c) $f = v/\lambda$ [1] = 340 m s^{-1} × 0.15 m [1] = 2.27 × 10^3 Hz [1].

2 (a) a represents acceleration [1], k represents the spring constant [1], x represents the displacement [1] and m represents the value of the mass [1].

(b) (i) 0.13 m [1].

(ii) $f = 1/T$ [1] = 1 ÷ 1.6 s = 0.625 Hz [1].

(c) $k = 4\pi^2 f^2 m$ [1]

$= 4\pi^2 \times (0.625)^2 \times 0.5$ [1]

$= 7.7$ N m^{-1} [1].

(d) (i) $v_{max} = 2\pi f A = 0.51$ m s^{-1} [1], E_k at this speed = ½ × 0.50 kg × (0.51 m s^{-1})^2 [1] = 6.5 × 10^{-2} J [1].

(ii) At zero displacement [1].

3 (a) (i) The minimum possible energy of an electron [1].

(ii) An electron that is just free has zero energy [1]. An electron in the ground state needs to gain energy to become free [1].

(b) (i) 12.75 eV × 1.60 × 10^{-19} J eV^{-1} [1] = 2.04 × 10^{-18} J [1].

(ii) $f = E/h$ [1] = 2.04 × 10^{-18} J ÷ 6.63 × 10^{-34} J s = 3.08 × 10^{15} Hz [1].

(iii) Ultraviolet [1].

(iv) Danger, any one from skin cancer, blindness [1]. Use, any one from security marking, fluorescent lights [1].

4 (a) (i) To produce two identical wave sources [1] by diffraction [1].

(ii) This is close to the size of the wavelength [1] to ensure adequate spreading [1].

(b) At a point equidistant from the gaps [1]. The waves are in phase [1] and have the greatest combined amplitude [1].

(c) Separation of maxima, $x = \lambda D/a$ [1] = 2.8 × 10^{-2} m × 1.20 m ÷ 0.10 m [1] = 0.336 m [1]. The distance between the strongest and weakest signals is half this, i.e. 0.168 m [1].

(d) The slits are many wavelengths wide for light [1] so no diffraction occurs [1].

5 (a) (i) hf is the energy of a photon [1], Φ is the work function, i.e. minimum photon energy that causes emission [1] and $(\frac{1}{2}mv^2)_{max}$ is the maximum amount of kinetic energy of an emitted electron.

(ii) Some electrons need more than the minimum energy to liberate them, so they have less than the maximum kinetic energy [1].

(b) (i) $E = hf = 2.02 \times 10^{15}$ Hz × 6.63 × 10^{-34} J s [1] = 1.34 × 10^{-18} J [1].

(ii) 4.5 × 10^{-6} W m^{-2} × 2.5 × 10^{-6} m^2 ÷ 1.34 × 10^{-18} J [1] = 8.40 × 10^6 [1].

(iii) $hf - (\frac{1}{2}mv^2)_{max}$ [1] = 1.34 × 10^{-18} J – 4.05 × 10^{-19} J = 1.30 × 10^{-18} J [1] ÷ 1.60 × 10^{-19} J eV^{-1} = 8.1 eV [1].

6 (a) $A = 0.28$ m [1], $f = 0.67$ Hz [1].

(b) $v_{max} = 2\pi f A = 1.18$ m s^{-1} [1].

(c) $v = 2\pi f \sqrt{(A^2 - x^2)}$ [1] = 0.82 m s^{-1} [1]

3 Fields

1 (a) $F = kQ_1Q_2/r^2$ [1] $= 9.0 \times 10^9$ N m^2 C^{-2} $\times (1.6 \times 10^{-19}$ C$)^2 \div (5.2 \times 10^{-11}$ m$)^2$ [1] $=$
8.5×10^{-8} N [1].

   (b) (i) $E = kQ/r^2$ [1] $= 9.0 \times 10^9$ N m^2 C^{-2} $\times 1.6 \times 10^{-19}$ C $\div (5.2 \times 10^{-11}$ m$)^2$ [1] $=$
5.3×10^{11} N C^{-1} [1].

      (ii) From the proton towards the electron [1].

   (c) $v = \sqrt{(r \times F \div m_e)}$ [1] $= \sqrt{(5.2 \times 10^{-1}}$ m $\times 8.5 \times 10^{-8}$ N $\div 9.1 \times 10^{-31}$ kg) [1] $=$
2.2×10^6 m s^{-1} [1].

2 (a) (i) 1250 V [1]

      (ii) $E = qV$ [1] $= 10 \times 1.6 \times 10^{-19}$ C $\times 1250$ V [1] $= 2.0 \times 10^{-15}$ J [1].

   (b) 4.0×10^{-15} J [1].

   (c) (i) Top plate positive and lower plate negative [1].

      (ii) $Vq/d = mg$ [1] $V = mgd/q = 5.0 \times 10^{-5}$ kg $\times 10.0$ m s^{-2} $\times 6.0 \times 10^{-3}$ m $\div 1.6 \times$
10^{-18} C [1] $= 1.9 \times 10^{12}$ V [1].

3 (a) (i) $g = GM_E/r^2$ [1] $= 6.7 \times 10^{-11}$ N m^2 kg^{-2} $\times 6.0 \times 10^{24}$ kg $\div (3.8 \times 10^8$ m$)^2$ [1] $=$
2.8×10^{-3} N kg^{-1} [1].

      (ii) 2.8×10^{-3} N kg^{-1} $\times 7.4 \times 10^{21}$ kg [1] $= 2.1 \times 10^{19}$ N [1].

      (iii) $T = 2\pi\sqrt{(rM_M/F)}$ [1] $= 2\pi\sqrt{(3.8 \times 10^8}$ m $\times 7.4 \times 10^{21}$ kg $\div 2.1 \times 10^{19}$ N) [1] $=$
2.3×10^6 s [1].

   (b) (i) $F = GMm/r^2$ [1] $= 6.7 \times 10^{-11}$ N m^2 kg^{-2} $\times 7.4 \times 10^{21}$ kg $\times 1.0$ kg $\div (3.8 \times$
10^8 m$)^2$ [1] $= 3.4 \times 10^{-6}$ N [1].

      (ii) The Sun has a greater mass than the Moon [1] but it is very much further
away from the Earth [1] so its gravitational field strength at the surface of
the Earth is less [1].

      (iii) In position A the gravitational pulls of the Moon and the Sun are in the
same direction [1], in position B they are in opposite directions, so the
resultant force on the water is smaller [1].

4 (a) From N to S (radially outwards) [1].

   (b) Out of the paper [1].

   (c) The direction of the force on the electromagnet reverses [1] when the current
reverses direction [1].

   (d) $F = BIl = BI2\pi rN$ [1] $= 0.85$ T $\times 0.055$ A $\times 2 \times \pi \times 2.5 \times 10^{-2}$ m $\times 150$ [1] $= 1.1$ N [1].

5 (a) (i) $E_k = eV$ [1] $= 1.6 \times 10^{-19}$ C $\times 2500$ V [1] $= 4.0 \times 10^{-16}$ J [1].

      (ii) $v = \sqrt{(2E_k/m)}$ [1] $= \sqrt{(2 \times 4.0 \times 10^{-16}}$ J $\div 9.1 \times 10^{-31}$ kg) [1] $= 3.0 \times 10^7$ m s^{-1} [1].

   (b) (i) Down (from top to bottom of the paper) [1].

      (ii) There is an unbalanced force on the electrons [1] which is always at right
angles to the direction of motion [1].

      (iii) $r = mv/Bq$ [1] $= 9.1 \times 10^{-31}$ kg $\times 3.0 \times 10^7$ m s^{-1} $\div (2.0 \times 10^{-3}$ T $\times 1.6 \times 10^{-19}$ C)
[1] $= 8.4 \times 10^{-2}$ m

6 (a) (i) $\phi = BA$ [1] $= 0.15$ T $\times 8.0 \times 10^{-2}$ m $\times 5.0 \times 10^{-2}$ m [1] $= 6.0 \times 10^{-4}$ Wb [1].

      (ii) 0 [1].

   (b) (i) The reading is a maximum when the plane of the coil passes through the
horizontal position [1] as the rate of change of flux is greatest [1]. The
reading decreases to zero when the plane of the coil is horizontal [1] as the
rate of change of flux decreases to zero [1].

      (ii) Generator/dynamo [1].

   (c) Any three from: increase the speed of rotation, increase the number of turns on
the coil, increase the area of the coil, increase the strength of the magnetic field
[1 mark each].

4 Particle physics

1 (a) 298 K [1].
　(b) $n = pV/RT$ [1] = 2.02 × 10^5 Pa × 2.20 × 10^{-6} m^3 ÷ (8.3 J mol^{-1} K^{-1} × 298 K) [1] = 1.80 × 10^{-4} [1].
　(c) (i)　The total energy of the particles [1].
　　 (ii)　Kinetic [1].
　　 (iii) It doubles [1] as internal energy is directly proportional to absolute temperature [1].
　　 (iv) At constant pressure, the volume is directly proportional to absolute temperature [1].
　　　　 The new volume = 2 × 2.20 × 10^{-6} m^3 = 4.40 × 10^{-6} m^3 [1].

2 (a) (i)　$^{239}_{92}$U [1]
　　 (ii)　The nucleus splits [1] into two smaller nuclei [1] and spare neutrons [1].
　　 (iii) The neutrons can cause other nuclei to fission [1] which releases neutrons to cause more fissions [1].
　(b) (i)　The total mass-energy of the products [1] is less than that of the original nucleus and neutron [1].
　　 (ii)　Kinetic energy [1] of the fission products [1].
　　 (iii) The energy is removed by a coolant [1] and used to generate steam [1] which drives a steam turbine [1].
　(c) (i)　The control rods absorb excess neutrons [1].
　　 (ii)　The moderator slows down the neutrons [1].

3 (a) $t = mc\Delta\theta/P$ [1] = 1.50 kg × 4.2 × 10^3 J kg^{-1} K^{-1} × 90 K ÷ 2.40 × 10^3 W [1] = 236 s [1].
　(b) The element absorbs energy [1], the casing absorbs energy [1], energy is lost through conduction/convection/evaporation/radiation [1].

4 (a) Gamma [1], gamma can penetrate tissue but beta is absorbed by tissue [1].
　(b) The beta radiation emitted by iodine-131 is absorbed and causes ionisation [1], resulting in the damage of cells/tissue [1].
　(c) Iodine-123 has a shorter half-life [1] so less is needed to produce the same rate of decay [1].
　(d) After 6 half-lives [1] = 78 hours [1].

5 (a) (i)　Electrons are not repelled by the nucleus [1]. They have wave-like behaviour so are diffracted by a nucleus [1].
　　 (ii)　The higher the energy, the shorter the wavelength [1]. The wavelength needs to be short so that it is of the order of the size of a nucleus [1] for diffraction to occur [1].
　(b) (i)　R = 1.20 × 10^{-15} m × 4$^{1/3}$ [1] = 1.90 × 10^{-15} m [1].
　　 (ii)　3.81 × 10^{-15} m [1].
　(c) The densities are the same [1]. The radius of the sulphur nucleus is twice that of the helium nucleus [1] so its mass should be 2^3 = 8 times as great if it has the same density, which it is [1].

6 (a) This shows that most alpha particles do not go close to a region of charge [1], so most of the volume of an atom is empty space [1].
　(b) There is a tiny concentration of charge in the atom [1] which must be the same charge as the alpha particles [1] to repel them [1]

5 Medical physics

1 (a) The intensity decreases [1].

(b) (i) Intensity = 10 log (I/I_0) [1] = 10 log $(6.0 \times 10^{-5}$ W m^{-2} ÷ 1.0×10^{-12} W m$^{-2})$ [1] = 78 dB [1].

(ii) $r_0 = \sqrt{r_1^2(I/I_0)}$ [1] = $\sqrt{(6.0 \times 10^{-5}$ W m^{-2} ÷ 1.0×10^{-12} W m$^{-2})}$ [1] = 7.7×10^3 m [1].

(iii) 0 [1].

(c) The perceived loudness of a sound is greatest at a frequency of 3000 Hz [1]. It decreases [1] as the frequency becomes lower or higher [1].

2 (a) An A-scan [1] as this measures depth below the surface of the skin [1].

(b) $t = 2 \times s ÷ v$ [1] = 0.12 m ÷ 1500 m s^{-1} = 8.0×10^{-5} s [1].

(c) So that there is no interference [1] between the pulse and its reflection [1].

(d) There needs to be a time lag when the crystal is not emitting ultrasound [1] so that it can detect the reflection [1].

(e) $\lambda = 6.0 \times 10^{-2}$ m ÷ 200 = 3.0×10^{-4} m [1]; $f = v/\lambda$ = 1500 m s^{-1} ÷ 3.0×10^{-4} m [1] = 5.0×10^6 Hz [1].

(f) The gel maximises the energy transmitted into the body [1] by reducing the amount that is reflected [1].

(g) (i) A CAT-scanner uses an X-ray beam [1] that rotates around the body [1] and produces a three-dimensional image of the body [1].

(ii) Advantages: the CAT-scan produces a three-dimensional image [1], tumours are more easily detected [1]. Disadvantage: X-rays can be more damaging to body cells and tissue [1].

6 Astronomy and cosmology

1 (a) (i) Apparent magnitude is a measure of the brightness when viewed from the Earth [1].

Absolute magnitude is the apparent magnitude the star would have at a distance of 10 pc [1].

(ii) The perceived brightness of a star decreases with increasing distance [1]. Absolute magnitude takes account of this [1].

(b) $M = m - 5$ log $(d/10)$ [1] = 1.3 [1].

(c) $d = 10$ log^{-1} $(m-M)/5$ [1] = 240 pc.

(d) White dwarfs are smaller than red giants [1] so radiation is emitted from less surface area [1].

2 (a) An absorption spectrum is what remains of the full spectrum after passing through an element [1]. Elements absorb the same wavelengths as they emit [1] so the absorption spectrum has these wavelengths removed from the full spectrum, leaving dark lines [1].

(b) (i) The star is moving away from the Earth [1], causing the wavelengths received to be longer than those emitted [1].

(ii) $v = c\Delta\lambda/\lambda$ [1] = 3.00×10^8 m s^{-1} × (0.277 ÷ 6.563) [1] = 1.27×10^7 m s^{-1} [1].

(iii) $d = v/H$ [1] = 5.78×10^{24} m [1] to 6.18×10^{24} m [1].

3 A shows an open Universe [1], the Universe will continue to expand if there is insufficient mass for gravitational forces to stop this [1].

B shows a steady-state [1] where the expansion stops due to gravitational attraction of the mass in the Universe, for this to happen the Universe needs to have the critical mass [1].

C shows the Universe contracting to a 'Big Crunch' [1], this will only happen if the mass of the Universe is greater than the critical mass [1]

7 Synoptic assessment

1 (a) To produce a parallel beam [1].
   (b) (i) from right to left [1]
       (ii) from left to right [1]
       (iii) into the paper [1].
   (c) To move in a straight line, $Bqv = Eq$ [1] so $v = E/B$ [1].
   (d) (i) The only force acting is the magnetic force [1], this is always at right angles to the direction of motion [1].
       (ii) $r = mv/Bq$ [1] $= 2.0 \times 10^{-26}$ kg $\times 2.4 \times 10^{5}$ m s^{-1} ÷ (0.15 T $\times 1.6 \times 10^{-19}$ C) [1] $= 0.20$ m [1].
   (e) The ratio of carbon-14 atoms to carbon-12 atoms decreases after a tree is felled. Use a sample of carbon from the wood in the table in the spectrometer; the carbon-14 atoms follow a path with a greater radius of curvature than that of the carbon-12 atoms [1]. The ratio of carbon-14 to carbon-12 can be determined by using counters [1].

2 (a) $C = Q/V$ [1] $= 4.8 \times 10^{-6}$ C $\times 3.6 \times 10^{5}$ V [1] $= 1.33 \times 10^{-11}$ F [1].
   (b) $W = \frac{1}{2}QV$ [1] $= \frac{1}{2} \times 4.0 \times 10^{-6}$ C $\times 3.0 \times 10^{5}$ V [1] $= 0.60$ J [1].
   (c) (i) $Q = Er^2/k$ [1] $= 3.0 \times 10^{6}$ N C^{-1} $\times (0.12$ m$)^2$ ÷ 9.0×10^{9} N m^2 C^{-2} [1] $= 4.8 \times 10^{-6}$ C [1].
       (ii) $V = kQ/r = Er$ [1] $= 3.0 \times 10^{6}$ N V^{-1} $\times 0.12$ m [1] $= 3.6 \times 10^{5}$ V [1].
   (d) (i) $F = Eq$ [1] $= 3.0 \times 10^{6}$ N C^{-1} $\times 1.6 \times 10^{-19}$ C [1] $= 4.8 \times 10^{-13}$ N [1].
       (ii) The force on the outer electrons ionises the air [1], movement of these ions causes further ionisation [1], the dome discharges as it attracts ions of the opposite charge to it [1].

3 (a) Collisions with meteors [1].
   (b) Any two from traffic, wind, tides [2].
   (c) The mean speed of the Moon is calculated by assuming that the path is circular [1] with a mean radius of 3.81×10^{8} m [1].
   mean speed $= (2 \times \pi \times 3.81 \times 10^{8}$ m$)$ ÷ $(27.3 \times 24 \times 60 \times 60$ s$)$ [1]
   $= 1.01 \times 10^{3}$ m s^{-1} [1].
   The gravitational force between the Moon and the Sun is the centripetal force [1].
   Force at closest distance $= mv2/r$ [1]
   $= 7.4 \times 10^{22}$ kg $\times (1.01 \times 10^{3}$ m s$^{-1})^2$ ÷ $(3.56 \times 10^{8}$ m$)$ [1]
   $= 2.1 \times 10^{20}$ N [1]
   Force at greatest distance $= 1.9 \times 10^{20}$ N [1].
   The difference between these forces $= 2 \times 10^{19}$ N.

   The actual difference is likely to be greater than this because of the assumption that the Moon travels at a constant speed [1]. In fact it speeds up as it gets closer to the Earth and slows down as it moves further away [1]. The effect of this would be to make the largest force larger and the smallest force smaller [1].

Index